Mary

ALSO AVAILABLE FROM BLOOMSBURY

What Is Philosophy For? Mary Midgley
Can't We Make Moral Judgements? Mary Midgley
The History of Animals: A Philosophy, Oxana Timofeeva

Mary Midgley

An Introduction

GREGORY S. MCELWAIN

BLOOMSBURY ACADEMIC

LONDON • NEW YORK • OXFORD • NEW DELHI • SYDNEY

BLOOMSBURY ACADEMIC
Bloomsbury Publishing Plc
50 Bedford Square, London, WC1B 3DP, UK
1385 Broadway, New York, NY 10018, USA

BLOOMSBURY, BLOOMSBURY ACADEMIC and the Diana logo are trademarks
of Bloomsbury Publishing Plc

First published in Great Britain 2020

Cover design by Maria Rajka
Cover image: Cinereous Owl from John James Audubon's *The Birds of America*
(© Granger / Bridgeman Images)

A catalogue record for this book is available from the British Library.

A catalog record for this book is available from the Library of Congress.

ISBN: HB: 978-1-3500-4757-0
PB: 978-1-3500-4756-3
ePDF: 978-1-3500-4758-7
eBook: 978-1-3500-4759-4

Typeset by Deanta Global Publishing Services, Chennai, India
Printed and bound in Great Britain

To find out more about our authors and books visit www.bloomsbury.com
and sign up for our newsletters.

To the memory of Mary Midgley (1919–2018)

Contents

Acknowledgments

This book would not have been possible without the support of many people along the way. I'd first like to thank the whole team at Bloomsbury for their help and support in bringing this book to fruition. I'm also grateful to the many friends and colleagues who generously provided feedback on various chapters of the book: Amy Brown, Luke Coddington, Katie Devine, Ian J. Kidd, Lucas Johnston, Todd LeVasseur, Benjamin Lipscomb, Kendall Marchman, Liz McKinnell, Anna Peterson, Holmes Rolston III, Elizabeth Wakeman, Jen Wallin-Ruschman, and Rachael Wiseman. This interdisciplinary collection of scholars (whose specialties include gender, neuroscience, astrophysics, philosophy of religion, sustainability studies, environmental studies, ethical theory, religious studies, political philosophy, animal ethics, religious ethics, environmental philosophy, feminist philosophy, community psychology, and philosophy of mind, among others) is a testament to the expansive scope and relevance of Midgley's thought. I am also thankful for the help of Skylar Barsanti and Max Schaffner, who read, edited, and provided feedback on the entire manuscript. Thanks also are due to the anonymous reviewer who provided feedback on the initial draft of the manuscript for Bloomsbury.

One of the highlights of this project, which deserves special mention, was to receive contributions and input from the late Mary Warnock (1924–2019). It has been a real pleasure interacting with such inimitable philosophical legends.

I must acknowledge Taylor & Francis and Les ateliers de l'éthique/The Ethics Forum for permissions to rework portions of two publications into this volume, which appeared previously as Gregory S. McElwain, "The Mixed Community," in *Science and the Self: Animals, Evolution, and Ethics: Essays in Honour of Mary Midgley*, eds. Ian James Kidd and Liz McKinnell, 41–51 (2015), and Gregory S. McElwain, "Midgley at the Intersection of Animal and Environmental Ethics," Les Ateliers de l'Éthique/the Ethics Forum

13, no. 1 (2018): 143–58. The latter was published under a creative commons license, Attribution 4.0 International (CC BY 4.0) (https://creativecommons.org/licenses/by/4.0/legalcode). This project was, furthermore, generously supported over the years by funding from the Faculty Growth and Development and the National Endowment for the Humanities funds at The College of Idaho. I have also received tremendous support from my colleagues in the Philosophy and Religious Studies Department at The College of Idaho.

I am, naturally, deeply grateful for Mary Midgley's time and accessibility over the years, and for permissions to use portions of our yearly interviews from 2011 to 2018 in the book. I am also honored by her work on the foreword to this book, which was never completed. Midgley largely influenced my intellectual and professional direction, and it has been a truly wonderful process getting to know her over the years. I can only hope this volume does justice to such a profound and prescient thinker. She was a mentor and an inspiration, and I will always be grateful for our visits in Newcastle.

And, finally, words cannot express my love and gratitude for Alyson, Emily, and my entire family.

1

Philosophical Plumbing

How to think involves how to live.[1]

Mary Midgley once argued that, among other things, philosophy is like plumbing. Like plumbing, our thought is supported by a complex underlying structure. And, like plumbing, this structure or network largely goes unnoticed—we rely on it for essential tasks and needs, and it more or less supports our everyday lives. That is, until something goes wrong. Overlooked thought patterns, like faults in the pipework, can slowly undermine the integrity of the entire system. These faults can be hard to repair, as they are often hard to locate and buried beneath the surface as part of a network that was never "consciously planned as a whole."[2] Typically the result of years of neglect, these faults require repairs if we are to carry on with life. Our thought patterns, in other words, are essential to our lives and require maintenance and repair to ensure their proper functioning. Midgley argues that, just as plumbers specialize in care and maintenance of drains and pipes, philosophers specialize in the care and maintenance of our thought patterns.

Midgley's plumbing metaphor is about the necessity of philosophy. Philosophy is, indeed, "grand and elegant," and worthy of pursuing as an end in itself. This she agrees. Yet, philosophy must remain grounded in the wider context of life, lest it become isolated and removed—something of a venture separated and independent from the rest of life. This independence can be perceived as intrinsic to the practice, as part of how we should regard the status of philosophy: something valuable but only done by intelligent, detached (and privileged) faculty in ivory towers. Philosophy can thus become regarded as a luxury, something interesting and valuable but not necessarily useful or necessary. Midgley resists this isolationism. She reminds us that philosophy is, among other things, largely practical. It involves how to think, and "how to think involves how to live."[3]

To stick with the plumbing metaphor, we may not always notice when something goes wrong with our thoughts and concepts. At least not initially. But as with plumbing problems, we tend to notice that something is off, "that life is going badly."[4] Something may go awry in our relationships or actions. We may start to feel increasingly isolated from each other or become alarmed of our treatment of others, including animals and the natural world. We may recognize asymmetries between groups or strange applications of scientific thought. It is not always clear how things got this way, but something seems out of sync in our lives. This, Midgley argues, is what Socrates was so valuable in noticing—that there are "limits to living in a mess."[5] Working our way out of this mess, insofar as is possible, involves reflecting on the many concepts and ways of thinking that underlie our everyday lives. Reflection of this sort is no easy process, as it requires that we "bend thought round so that it looks critically at itself."[6] Yet, this careful and meticulous work is indispensable if we are to find ways of dealing with the problems and conflicts that surface in our lives.

"Philosophy," Midgley writes, "like speaking prose, is something we have to do all our lives, well or badly, whether we notice it or not. What usually forces us to notice it is conflict."[7] All sorts of conflicts arise throughout our lives and require various types of solutions. On the whole, this process involves patching the leaks and clearing the blockages in our conceptual systems. We must interrogate our assumptions, which are often "muddled and inarticulate," and get our hands dirty to find what plagues our thought. Philosophy helps us examine and, when necessary, readjust our concepts. It deals in connecting concepts with wider elements of our thought and lives in coherent ways, in organizing concepts into "usable schemes."[8] These conceptual schemes, which are many, are largely the province of philosophy.

When our concepts go awry—as with, for instance, the runaway individualism of contract thought—philosophers are, in turn, the specialists to call. Conceptual repair is urgent business, as faulty concepts and visions of the world "quietly distort and obstruct our thinking" and, in turn, influence everyday life.[9] Though people are typically more aware of the need for plumbers than for philosophers, both are vital in examining and servicing the vital underlying structures of our lives. Philosophy is thus not a luxury but a remedy for faulty thinking. And, since our thought is interconnected, philosophy, writ large, is valuable. Ideas and concepts flow back and forth between us and affect us all. Sexist and anthropocentric (human-centered) patterns of thought, for instance, have real impacts on how we think of ourselves and treat each other and the natural world. Social contract thinking, with its assumptions of freely negotiated and annulled relationships, has real impacts on the way we see our social relations.[10] Scientistic thinking has real impacts on the perception of science. Philosophers, acquainted with the wider background of conceptual developments over time, provide a valuable service. They are able to sort through various "conceptual tangles" and diagnose problems in patterns of thought.[11] They help us to think more clearly.

Thus, Midgley argues, philosophers help us sort through the complex concepts and patterns of thought that ultimately support and influence our everyday lives. In this way, philosophy is a practical necessity. But it also plays a visionary role. Philosophy is not just the logical task of analyzing concepts and ideas. Though this difficult and disciplined work is essential in clarifying the conceptual tangles that hamper our thoughts and lives, it is not the whole of philosophy. Philosophy contains a visionary aspect that provides imaginative direction for life. "Great philosophers," Midgley writes, "need a combination of gifts that is rare. They must have both the new vision that points the way we are to go and the logical doggedness that sorts out just what is, and what is not, involved in going there."[12] Here she argues that this balance of skills in philosophers is like the balance between a lawyer and a poet. Too much "lawyer-like" reliance on the formal and technical elements in philosophy can result in esoteric scholasticism, while excessive focus on poetic visions without disciplined, detailed thinking can result in fantasy and idealism. The challenge, and the promise of philosophy, is in this delicate balance of critical analysis and visionary direction.

A philosophy for life

Though Midgley was far too modest to ever claim the status of "great philosopher," she clearly and accessibly exhibits the rare combination of skills and talents required of great philosophers. As with most great philosophers, she not only helps us work through various conceptual tangles but also offers visions of the way forward and direction on how to get there. For Midgley, this often involves working through the "hidden system of ideas" that lie underneath our notions of human nature and our connection to each other and the natural world.[13] These ideas and patterns of thought, she argues, have an interactive relationship with our myths, which are "imaginative networks of powerful symbols that suggest particular ways of interpreting the world."[14] Patterns of thought give rise to myths, and myths, in turn, influence the way we think.

Midgley emphasizes this dynamic relationship because myths (or "world-pictures") greatly impact how we live. Contract myths, creation myths, and myths of the omnicompetence of science all carry certain implications with them. This makes sense, as myths have strong symbolic power and motivational force. But their influence is dependent on their emphases. Myths are always partial and provisional pictures of the world. As beings that seek an integrated life, we often try to unify our thought under one overarching vision or narrative. She warns against this temptation. Though most myths convey some valuable elements in our efforts to make sense of the world—the importance of rational choice and autonomy (contract myths), our nature and origins (creation myths), and the explanatory power of science (omnicompetence myths)—they must connect with life as a whole, lest they become exaggerated and, ultimately, distorted and damaging. Midgley's philosophy is one that works through the patterns in our thought and how they interconnect with our mythic visions of the world. Though she carefully critiques and dissects particular arguments, one of her more distinctive contributions to philosophy is, in her words, "pointing out the wider imaginative landscapes that have made [the arguments] look plausible, the visions that shape the thought behind them."[15] Through "mapping" thought patterns and visions in this way, she offers guidance "out of various dead ends in which people easily get trapped."[16]

I explore some of the distinctive features of Midgley's mapping project throughout this book. For now, it will be helpful to preview her characteristic approach to the topics that arise in the following chapters. This blueprint of her approach is in broad strokes but represents the rough structure of her typical angle of attack on various concepts and visions. Since Midgley argues that philosophy is practical, any idea, whether in philosophy, science, or wider circulation among the public, is fair game. Very often, she laments, the ideas and concepts that have the strongest grip on people's imaginations are those that philosophers often overlook. Everyday thinking, Midgley argues, must be criticized, examined, and taken seriously, "even in its most lurid or odious manifestations. It needs to be noticed as thought and not just as noise."[17] She takes seriously all manner of ideas and visions because they have real impacts, whatever their origin. In turn, these ideas and visions can be interrogated, and our direction can be recalibrated. "By seeing what not to think," she argues, "we can often move towards the parts of the map which will help us. . . . The appeal of certain mistakes often lights up aspects of the problem which we would otherwise miss."[18]

Thus, Midgley identifies issues that are important and have real-world implications. They are often, moreover, ideas that represent the exaggerated extremes of a topic. This is because, whether in philosophy, science, or popular works, the extremes typically are the loudest voices in the debate and tend to have the widest impact on our thought and imagination. These positions also tend to operate in black-or-white terms, promoting the false dilemma that there are only "two positions: the right one and the other."[19] It is not that these extreme and intransigent positions have nothing to offer. Whether it be the nature-nurture debate or a clash between domestic animals and the environment, Midgley argues that extreme positions often do reflect important details or considerations that must be taken seriously. Accordingly, rather than all-purpose solutions, the extremes more typically contain "half-truths," or "one-sided proposals with a useful aspect which needs to be balanced by their other halves and then integrated into a wider framework."[20] While this competitive "tribalism" of ideas can draw out useful critiques and ideas at each end, it tends to oversimplify the issues and frames the debate in black-or-white terms. Midgley resists the tribal warfare model of ideas. She suggests, rather, that ideas and visions might better flourish in a more cooperative and nurturing context, invoking images of exploration and gardening over competitive dueling alone (though Midgley was never one to shy away from a good fight when necessary). The competition of ideas can bring out important aspects of a problem, but this practice only gets us so far. It can encourage a winner-takes-all mentality of ideas, favoring conflict and controversy over cooperation and careful attention to the important landscape between positions.

Midgley's "occupation," as she termed it, often involved interrogating the extremes in order to explore and map the important territory in between.[21] This approach does not mean that the answers are simply "in the middle." While more fruitful territory may indeed be somewhere near the middle, the most hospitable territory to our thought may be more toward one pole than the other—for instance, more toward realism than skepticism in science or more toward pluralism than monism in morality. Or, the answers may lie in some territory that has not yet been explored. This type of work was modeled by Aristotle, who looked at "the existing forms of thought to see what can be made of them before launching something quite new."[22] For Midgley, there is likely no real "golden mean" that solves all of our problems. But the answers, if they are to be found, are most often not in the extremes.

This balancing approach—if the results are to be meaningful—is more difficult than it may at first appear. Something more than a simple "it's in the middle" is necessary, and one must "dig deeper in order to bring together the various half-truths that were involved—to find a synthesis that did not just paper over the cracks."[23] One must seek complexity over simplification, compromise over conquest. For those engaged in the militaristic warfare of ideas, this position may seem too noncommittal—combatants must choose a side and defend it to the death. This mentality, however, often assumes that there is an endpoint where victory can be declared for one's given favored system of ideas. Midgley argues that the history of moral philosophy tells us otherwise:

> What great philosophers do for us is not to hand out such an all-purpose system. It is to light up and clarify some special aspect of life, to supply conceptual tools which will do a certain necessary kind of work. Wide though that area of work may be, it is never the whole, and all ideas lose their proper power when they are used out of their appropriate context. That is why one great philosopher does not necessarily displace another, why there is room for all of them and a great many more whom we do not have yet. Because our aims are not simple, we are forced somehow to reconcile many complementary principles and duties.[24]

Midgley's pluralistic understanding of ideas, in morality, science, or otherwise, does not assume that all ideas will eventually converge or that a great, overarching synthesis will emerge from the battle of thesis versus antithesis. Rather, she argues that "the whole truth is complicated and needs many [ideas]."[25] Things are too complex to reduce to one principle, truth, or simple set of ideas. This does not mean that all ideas or positions are created equally— some really are more valuable or valid than others and reasons can be given for why. This analytical work is part of the job of philosophy. In mapping concepts and visions in this way—as partial and never complete—Midgley sees the world as "containing a much wider spectrum of values, a world so complex that we need to look at it from an indefinite number of different angles."[26] This reflects Midgley's characteristic assertion that "life," after all, "is essentially messy."[27]

Doing philosophy in this way requires great dexterity in navigating some of the pitfalls that surface when developing ideas—for instance, the seductive notion that, if followed through doggedly, one's prized concepts or principles will save the day, as it were. It also requires that one resist the temptation to sensationalize things, a habit that often occurs in the public dissemination of ideas. (The nature-nurture debate of the 1970s–80s is a good example of this.) A certain level of shock value no doubt gets people's attention, but Midgley remarks, "If you shock people too much they can't hear you."[28] On the other hand, getting through to people requires a healthy dose of lively language and compelling ideas. Otherwise, she reflects, one can come across as "so cautious and qualified that it doesn't get across anybody."[29] Midgley's philosophical style, refined in its clarity and wit through many disputes, reflects this delicate balance: she is sharply critical of extreme positions and visions, cautious and incisive in working through conceptual tangles, and clear and compelling in giving directions forward out of a given problem. Some problems may only have partial resolutions, and others none. The point, though, is that there are better and worse ways of thinking through the issues that impact our lives.

Midgley's style is, furthermore, accessible and sensible. Her approach to complex issues in everyday language is no doubt part of why she is so widely read outside of professional philosophy. Jane Goodall nicely summarizes this accessibility:

> One of the reasons why I was, from the first, so impressed by her books was not just because of the issues she discussed but because, unlike many philosophers—and many scientists—her writing is not only elegant, but beautifully free of jargon. She can discuss a most complex and profound topic in such a way that her argument is immediately understandable.[30]

This accessible style was deliberate because Midgley saw philosophy as something that we all share. As such, it should be readable, practical, and, just maybe, even enjoyable at times. Accordingly, many of her contributions to philosophy reflect her lucid, witty, and common-sense bent. Of the many critiques and visions that she is known for, Midgley herself believed they were fairly straightforward. She lightheartedly reflected that "all my books, when I glance at them, seem terribly *obvious* [laughs]. At the time, I think, well, this is too obvious to say, but obvious things aren't said in an academic context."[31] What was "obvious" to Midgley, however, was not always quite so for others, as things do get lost in the process of forging and disseminating ideas. It takes a rare mind to recognize and work through the many intricacies of our concepts and visions in the skillful way that Midgley did for well over forty years, and this book represents an attempt to explore and interpret these contributions.

A life in philosophy

Before moving on to the goals and structure of the book, an overview of Midgley's life is in order. Much has been written about Midgley's life, most notably in her memoir, *The Owl of Minerva*, as well as in other volumes dedicated to her work.[32] For this reason, and because this book is an introduction to her philosophy, I must point readers toward those more biographical works while presenting a brief sketch here.

Mary Beatrice Scrutton Midgley was born on September 13, 1919, in London to Lesley (Hay) and Tom Scrutton. She spent her early life in Cambridge (where her father was chaplain at King's College, Cambridge), Greenford, and Ealing. Midgley recalls her childhood as one filled with curiosity and exploration in which her passion for wide learning developed. This, perhaps, was an early contribution to her profound interdisciplinary interests and acumen. She was educated at Downe House School—where she discovered ancient Greek philosophy and the significance of the history of ideas (and likely an early appreciation for the work of Charles Darwin)—and later at Somerville College, Oxford, where she studied Classics. It was there, during the Second World War, that she befriended Elizabeth Anscombe, Philippa Foot, and Iris Murdoch. This exceptional cohort of philosophers—recently dubbed "The Quartet"—has received considerable attention in recent years for collectively contributing to the transformation of British philosophy in the twentieth century, each with their own distinctive critiques and insights.[33] After graduating from Oxford, Midgley began work on a D.Phil. on Plotinus's doctrine of the self but never completed it. She attributes this to several issues, including the ambitiousness and breadth of the project along with challenges in operating the intricacies and rituals of Oxford philosophy. These challenges, along with what felt like a stifling and combative environment for philosophy, led her to leave Oxford in pursuit of other avenues. Had she remained, she writes, "I would certainly have got out of philosophy altogether."[34]

Fortunately, this fate did not transpire. Midgley took a position in philosophy at Reading University, which she found "liberating." She married the philosopher Geoff Midgley (who had worked under Gilbert Ryle) and moved to Newcastle. She had three children— David, Tom, and Martin—and began presenting radio and television broadcasts, reviewing books, and writing newspaper and magazine articles. These ventures were certainly indicative of her interest in the public and interactive nature of philosophy. In 1964 she was offered a position in the Department of Philosophy at Newcastle University. Midgley taught and flourished in this interdisciplinary environment until she retired in 1980. It was in Newcastle that she wrote her first book, *Beast and Man: The Roots of Human Nature*, in 1978, which, she writes, was "the trunk out of which all my various later ideas have branched."[35] Midgley was nearly sixty when *Beast and Man* was published, emblematic of her unorthodox path in philosophy. On this, she writes: "People often ask me why I did not start to write the books and articles for which I am now known until I was over fifty. The answer is simply that I needed time to think. Before then, I did not know what I thought clearly enough to want to go public with it."[36]

Since *Beast and Man*, Midgley authored an additional seventeen books, including her most recent publication *What Is Philosophy For?*, which was released during the month of her ninety-ninth birthday.[37] Over her career, she published over 250 articles, editorials, pamphlets, and other writings in addition to various broadcasts and other media (see "Bibliography"). She received wide recognition as a public intellectual, speaking at various festivals and events while writing for widely read publications (*The Guardian*, *New Scientist*, and *Philosophy Now*, to name a few). Midgley has been recognized many times over for her contributions to philosophy and was awarded an honorary D.Litt. by Durham University in 1995 (alongside Maya Angelou) and LLD from Newcastle University in 2008. She gave the prestigious Gifford Lectures at the University of Edinburgh in 1990 and received the Edinburgh Medal in 2015. In 2018, the Midgley Archives were dedicated at Durham University, and the Royal Institute of Philosophy's London Lecture Series was devoted to a centenary celebration of Midgley and the illustrious Quartet (Anscombe, Foot, Midgley, and Murdoch). Mary Midgley passed away on October 10, 2018, at the age of ninety-nine, a month after publishing her latest book (see "Afterword").

About the book

Though Midgley's philosophical career has come to an end, she will no doubt continue to excerpt an influence on the philosophical world. Her style, cogency, and range of coverage make her work a formidable presence in the field, though her appeal stretches beyond professional philosophy. Midgley has long been popular with students of philosophy and a wider audience trying to make sense of things—of human nature, morality, animals, nature, gender, science, religion—in integrated and realistic ways that resist oversimplification. I have written this introductory book primarily with these readers in mind, especially students. Midgley took students seriously, and I have tried to follow suit. (She requested, specifically, that the book be accessible in this way.) Thus, I make efforts to introduce terms and concepts as they appear and do not assume that the reader has an extensive background in any of the topics in this book. Moreover, I try to avoid relentless reference to other works and thinkers while providing mentions and context where necessary. These details and citations are mainly found in the endnotes.

Now on to the main objectives of the book. The first objective is, naturally, to introduce readers to Midgley's philosophy through the main topics, concepts, and visions to which she devoted her life's work. This involves bringing together a significant number of publications on a wide range of topics in one place. Given Midgley's prolific range of material and the introductory nature of the book, every aspect of her thought cannot be thoroughly examined in the manner it deserves. I've had to make some tough decisions on what to include and what to leave out due to the length and scope of this volume. Given my own specialties and interests, I have paid special attention to Midgley's animal and environmental thought, though one could certainly spotlight other areas of her corpus in similar ways. My hope, regardless, is that Midgley's main points get across and that interested readers are compelled to delve deeper into her work, which is far richer than can be captured here alone. An up-to-date bibliography of her work to date (thanks to the previous work of Andrew Gray, Ian James Kidd, Liz McKinnell, and David Midgley) can be found at the end of the book.[38]

The second objective is to emphasize the value and relevance of Midgley's philosophy, which lies both in her unique method (outlined above) and in her contributions to the particular topics in this book. She models a rare combination of gifts, deftly engaging complex concepts and ideas in philosophy while remaining clear and accessible enough to have influence in a number of areas outside of philosophy (anthropology, psychology, religious studies, etc.) and outside of academia altogether. Through this approach she demonstrates the relevance and importance of philosophy across disciplinary lines and in everyday life. And through her approach, she has made significant contributions, in varying degrees, to the topics covered in this book. I explore these contributions throughout, though not as deeply as a sustained look at any one of the topics (human nature, animals, gender, etc.) would entail. This said, my hope is that the reader walks away with an appreciation of the ways in which Midgley consistently and impressively works through a range of messy conceptual tangles while offering directions forward in our visions of the world. Ideally, students of philosophy will take her characteristic philosophical approach to heart. Simon Blackburn aptly captures her relevance: "Many young students sense well enough that in the present darkness, articulate and well-informed understandings of their scientific civilization, its values and politics is necessary. *They need their Midgleys.*"[39]

What are the conceptual tangles and visions that Midgley addresses in her work? They cannot all be explored in an introductory text, but I have loosely categorized her discussions of these ideas under the following topics: "Human Nature" (Chapter 2), "Morality" (Chapter 3), "Animals" (Chapter 4), "Nature" (Chapter 5), "Gender" (Chapter 6), "Science" (Chapter 7), and "Religion and Science" (Chapter 8). (I utilize these short titles in cross-references throughout the book.) These categories and topics are not meant to rigidly break up Midgley's thought into nice, tidy areas. Readers familiar with her work know this is *not* possible. Her thought is inherently connective, holistic, and expansive. Thus, all of these topics intersect and overlap throughout the book. For our purposes, the topical divisions are simply ways of collating and synthesizing her ideas on various interrelated issues, with the understanding that the barriers between them are permeable and fuzzy. This becomes clear from the first chapters, as Midgley's views of human nature underlie the entirety of the book, as does her critique of the gendered nature of philosophy, and so on. As such, the book is best seen as a whole. It is, in this way, structured sequentially, and each chapter builds on and integrates the previous ones. Chapters can be read in isolation, but the fullest picture (and conceptual/definitional scaffolding) comes from a sequential reading. And, if something important seems to be missing in one chapter, it is very likely in another (a side effect of bringing so much content and so many ideas together in one place).

Chapters 2 and 3 are quite broad and expansive. This is especially the case with Chapter 2, which explores conceptions of human nature and the self. I've tied a number of diverse topics together, however loosely, under the theme of integration or wholeness. The search for wholeness, or a more integrated conception of the self, is one of the more central themes in Midgley's work. She argues that human beings seek unity in the sense of being a *whole* or integrated person. We become fragmented or disintegrated when elements of the self are divided against each other. Much of Midgley's conception of human nature, in turn, arises from her critique of extreme and reductive positions that portray elements of human nature and the self as antithetical or oppositional: humans versus animals, freedom versus determinism, mind versus body, and so on. I outline Midgley's effort to reintegrate these components of the self as complementary, rather than oppositional, in our notions of human nature. This chapter, in turn, sets up her characteristic approach to various false antitheses that are critiqued in subsequent chapters: humans versus nature, male versus female, science versus religion, and so on. She goes to great lengths to show the type of complex and sophisticated synthesis that can arise by not indulging in extremist reductionism, which oversimplifies aspects or elements of the self into totalizing explanations of the whole. What develops is a sensible and highly influential understanding of the relational self in context. This notion of the self, as I'll trace, underlies her critique and ethics in the following chapters.

Chapter 3 continues to develop Midgley's conception of an integrated life, this time in relation to morality. For Midgley, morality can be conceived of as what is serious or important, and moral philosophy is the process of working through central issues in our lives. Morality also grows out of our nature, which is deeply social. As such, she argues that understanding this nature may help us in reconciling some of its many conflicts in pursuit of a more integrated life. However, this reflective process is far from simple or straightforward. Quite often, the difficult work of weighing and relating values and claims is oversimplified in an attempt to manage the complexity of moral life. In the process, important elements of life as a whole, such as relationships, can be neglected and portrayed as antithetical to the ideals of freedom. Midgley resists this fragmenting approach to morality much as she did in relation to human nature. She addresses these and other simplifications—such as moral skepticism—arguing that moral philosophy must involve the whole person in context. This chapter concludes by elaborating on the possibility of moral change in society. Cultures do have the resources to flourish, though these often require reexamination in the process—for instance, as when egalitarian reformers assumed that equality only meant equality for *men*. This chapter outlines Midgley's general approach to morality. I trace the application of this approach in the following chapters on animals, nature, and gender. On these topics, Midgley focuses especially on the ways that individualist ideals and categories of morality are particularly detrimental for animals, nature, and women.

Chapters 4 and 5 cover animals and nature, respectively, and trace Midgley's influence and thought in these areas. Animals are a central element in Midgley's thought. Her work, from the start, focuses on the centrality of *humans as animals* and the broader implications that grow out of this premise. Chief among her priorities is challenging the conceptual barriers that have been placed between humans and the natural world. This largely "destructive" work, as she calls it, is part of her project, which begins in Chapter 2, of grounding conceptions of human nature in the natural world. Humans are firmly embodied and embedded creatures in Midgley's philosophy, continuous with nature and its many inhabitants. Animals, in this context, are our fellow beings, part of a "mixed" human-animal community. As such, they are important subjects, rather than objects or things, connected to us through powerful emotional and relational bonds. I argue that her relational approach in the early 1980s was among the first in animal ethics to look outside of rationalist individualist categories of morality. By focusing on the emotive and communal aspects of our relation to animals, she provided a shift from conceiving of animals in purely individualist terms (e.g., as bearers of rights or interests), highlighting the dynamic interrelationality and interdependence that binds individuals and the various communities and collectives in which they are located. This connects with her approach to the environment more broadly, locating us within the wider holistic context of the biosphere, which is composed of whole ranges of individuals and collectives, all of which matter in varying measures. Her prominent concept of the mixed community envisions animals as part of our wider community on earth and central among the many parts that constitute the valuable whole of our lives.

Chapter 5 traces how Midgley's environmental thought is also centrally connected to her conception of human nature. As with the topic of animals, we find Midgley often engaged in the destructive work of challenging the barriers that have been placed between humans and the natural world. Here, the most restrictive barriers are attitudes and ideas of human separation from, and superiority over, nature. These anthropocentric notions abstract humans from the natural world for religious or philosophical reasons and marginalize the natural world as the mere stage for the human drama. This abstraction and separation, along with mechanistic views of the world as passive and inert matter, have justified the exploitative and careless treatment of nature. This is furthered by hyper-individualist ideals of freedom, which frame the world as a resource to be utilized by humans in the competitive pursuit of freedom. Midgley's environmental thought pushes back against these damaging myths, advocating more relational and holistic visions of our deep connection to nature. Her organic and holistic visions offer more comprehensive and cooperative ways of understanding our interdependence and interconnection with each other and the wider world. In turn, Midgley promotes a vision of the wholeness of the self as constituted by its many relations to the human and nonhuman world. The earth, in this vision, is not our enemy or a resource but rather our home.

Chapter 6 explores gender. Midgley had a career in philosophy that spanned all four "waves" of feminism. As such, she engaged a range of issues related to feminism, sex, and gender, though these topics were rarely her main focus. This said, Midgley, along with her collaborator Judith Hughes, drew attention to the connection of sex and gender with wider problems in the Western philosophical tradition. Central among these, again, are hyper-individualist approaches that fragment notions of human nature and the self. These approaches promote a hypermasculine ideal, elevating (presumed) "masculine" virtues and perpetuating false hierarchical antitheses—reason versus emotion, mind versus body, masculine versus feminine, and so on. Midgley argues that combating masculinism and this oppositional way of seeing the world should be among the main priorities for feminist movements and philosophy. Though feminism and philosophy have changed in dramatic ways since the early 1980s when Midgley and Hughes wrote their work, much of their critique of individualism, masculinism, and the gendered assumptions of philosophy remains relevant.

Chapters 7 and 8 transition away from ethical and political visions toward scientific and religious visions of the world (though these are all interconnected in her analysis). Science is a powerful lens onto the world, shaping the ways in which we conceive of reality while, at the same time, being shaped by the larger mythic structure of our lives. Midgley argues that if we attend to science's functions and limitations, then we can use it wisely and sensibly. If we do not, then we run the risk of misunderstanding and, even worse, misusing science. It is also important to understand its connection to other parts of life, including other forms of knowledge. For Midgley, ideologies such as scientism and reductionism can distort this understanding of science in context. Scientism excessively isolates and glorifies science while reductionism offers overly simplistic, one-sided explanations of the world. These ideologies typically neglect other important variables, both within and outside of science, that can inform and guide a more nuanced appreciation of science in context. This appreciation means taking seriously the insights of both realism and skepticism in conceptualizing science as a vital element in the pursuit of our best understanding of the world. Exaggerations of the role of science can lead us astray and provoke hostility toward science. Furthermore, these exaggerations can distort the perception of science by making claims and extrapolations that overstep the bounds of scientific inquiry. These near-prophetic exaltations and distortions of science, she contends, also contribute to an interrelated cluster of problems with implications on the relationship between religion and science. Specifically, they can further the narrative of the warfare of religion and science while, paradoxically, fulfilling many of the functions of religion. In this way, *scientism*, rather than science, can look like a religion.

Midgley argues that, given science's underlying motivational and mythic structure—which is, to a certain extent, shared with religion—it is no wonder that scientistic isolationism and glorification can *look* religious. Of course, this argument depends on what religion looks like, which is complicated. With this complexity in mind, Midgley embarks to disentangle some of the conceptual issues involved in the relationship between religion and science, which entails working through some of their similarities and differences. However, this examination does not amount to a project of merely tracing parallels or equating science to religion. Rather, it represents an attempt to maintain the integrity of each pursuit in the context of life, in which distortions of each threaten to morph into surreal and twisted versions of the other. These potential consequences are part of the complex and, at times, convoluted relationship between these powerful influences in human life. Midgley ultimately seeks to contextualize and, to an extent, reconcile the more damaging ways in which religion and science have been portrayed and related. In the process, she draws attention to their dual contributions to our visions of the world, and how these visions can go awry. Visions that claim to have all the answers, particularly in fundamentalist distortions of religion *and* science, deceive us and obscure the bigger picture. The finality of unbalanced myths belies the complexity of the world, as well as the variety of imaginative tools that are at our disposal in conceptualizing the world in its wholeness.

I conclude with a brief remembrance of Mary Midgley and spotlight her final work and concerns.

2

Human Nature and the Self

I started by insisting that we are animals and drawing the consequences from that that we are part of the natural world as a whole.[1]

The search for wholeness—a unified or integrated conception of the self—is one of the more central themes in Midgley's work. Wholeness is perhaps best understood here as an integrated life that fits together as a whole. It is probably unrealistic and misguided to expect our lives to unify into some final completeness—life is simply too messy and complex to be so tidily packaged. But this does not mean that the many diverse elements of human nature and the self cannot be brought together in sensible ways. We can become disintegrated or fragmented when these elements are isolated and divided against each other, and we need not go out of our way to exaggerate these divisions. This is precisely what happens when aspects of our world are divided into warring alternatives: biology versus culture, reason versus emotion, self versus others, and so on. For over forty years, Midgley endeavored to reintegrate these and other components of human life and the self as complementary, rather than oppositional. She long demonstrated the complex and sophisticated synthesis that can arise by not indulging in extremism and reductionism. What develops, in turn, is Midgley's influential conception of the integrated and relational self, a vision which stems from our nature as social animals in context.

"Human nature"

Human nature was in the spotlight when Midgley published her first book *Beast and Man* in 1978. This was, in large part, due to the prominence of the "nature versus nurture" debate of the 1970s–80s. Prevailing accounts, in this debate, often dramatized aspects of human nature in oppositional terms: behavior and personality were determined *either* by internal forces *or* external ones, by biology (genes) or society (environment). "Tabula rasa," or blank slate, accounts of human nature—that we are pure products of "culture, originally indeterminate items, infinitely malleable, dough or blank paper at birth, shaped only by education"[2]—clashed with accounts that emphasized the power of our genetic inheritance. While behaviorist psychologists focused almost entirely on external, observable behavior as the only scientific way to study humans, sociobiologists touted the internal sources of human behavior as comprehensive.[3] In the first case, inner motives are ignored or denied, while in the latter, motives are explained (and often explained away) as the products of genes. While sensible proponents on each side drew attention to the gray area in between, the rhetoric of the debate tended to drift toward the familiar extremes of nature *versus* nurture.

Other voices further complicated these notions involving human nature. Humanists and social scientists gravitated toward the position that we are so malleable as to have *no nature at all*. This was influenced by both religious and existentialist positions, which, though seemingly quite opposed to the views of the other, agreed that human life is not really a product of nature, but of spirit or free choice. This is characterized by Sartre's famous credo that "existence precedes essence."[4] In these positions, the possibilities of human life are seemingly limitless and defy the constraints of *both* nature and nurture. This "no nature" approach was taken further still by social reformers that resisted the very term "human nature," a notion which had previously been distorted to support objectionable racist, sexist, and classist ideologies. These critics (and Midgley) saw fatalistic claims about human nature—especially ones that emphasized our selfish, competitive, and egoistic natures—as dangerously close to those of Social Darwinism and other ideologies that promoted the superiority of some groups over others.[5] "Human nature" was thus denounced and abandoned by some of those wary of its historical misuses.

Midgley insists there are more balanced ways to approach human nature than the reductionist, sensationalist, and ideological ones that turned some away from the topic. She agrees with reformers that the term had been egregiously abused but diverges from the conclusion that human nature, as a concept, must therefore be abandoned altogether.[6] She argues that better or worse conceptions of human nature will remain, regardless. As self-reflective beings, we cannot help but think of who we are and what our existence entails. What is the human condition? Why do we behave in the ways that we do? Can we improve things? Conceptions of human nature naturally flow from this process and permeate our thought, underlying and influencing our notions of self, others, and how to act in the world. The focus, then, should be to make clear and scrutinize these conceptions in the light of day to correct their potential flaws and implications. She writes:

> This project of cutting out the whole idea of human nature is, I think, another striking example of an idea understandably taken up for strong and reputable moral reasons, but stretched to do much wider work beyond its capacity. It is a classic case of curing a cold by cutting off your head. The proper response to institutions stained with racism, sexism and other forms of political oppression is to say just what is wrong with them and to point out means by which they can in fact be changed.[7]

That is, if a concept is being misused, the wisest response is not to abandon it, but to use it properly and, insofar as possible, reconcile its damages. Effort should then be made to ensure that human nature is studied well, understood properly, and communicated with clarity and caution.[8] The lack of this caution in the context of the nature-nurture debate, in turn, gave Midgley an opportunity to work through some general issues and problems in philosophy. These issues and problems—many of which stretch beyond this chapter and are covered throughout the book—are rooted, to an extent, in the general human tendency to abstract, distance, or otherwise remove humans from our natural context.[9]

Humans as animals

Midgley's far-reaching account of human nature and the self begins with the oft-quoted line: "We are not just rather like animals; we *are* animals."[10] This emphasis on our animal nature is by no means new, as philosophers dating back to Aristotle have explored this central aspect of the human condition. Yet, for Midgley, this premise is particularly important and is the starting point for much of her philosophy. This is because concepts and ideas relating to human nature and the self—who we are, where we come from, what our possibilities and limits are—must be made intelligible in context. Humans are a particular species with particular needs and limitations. Though this premise may seem relatively straightforward and given, it does not always play out so directly. Theorists of the human condition, guided by a tantalizing vision of human life, often lose track of this natural context, pushing it so far into the background as to abstract us from the world. We can then be imagined to be virtually anything, akin to Sartre's declaration that

> man first of all exists, encounters himself, surges up in the world—and defines himself afterwards. If man as the existentialist sees him is not definable, it is because to begin with he is nothing. He will not be anything until later, and then he will be what he makes of himself. Thus, there is no human nature, because there is no God to have a conception of it. Man simply is. Not that he is simply what he conceives himself to be, but he is what he wills, and as he conceives himself after already existing—as he wills to be after that leap towards existence. Man is nothing else but that which he makes of himself.[11]

Visions of the detached intellect or rational will, free from the restraints and conditions imposed on other beings, appeal to those wishing to privilege and even glorify reason and free will. Though there is indeed much to praise about these human qualities, Midgley writes of how certain fantasies and myths of the human condition can risk losing touch with reality (and reason itself):

> We are not disembodied intelligences, tentatively considering possible incarnations. We have highly particular, sharply limited needs and possibilities already. . . . . There are quite narrow restrictions on what can possibly be rational for such a being— not necessarily limits to the possible particular desires, but to the policies, the schemes of life into which these can possibly be built. . . . Not every form of life can make sense for a given species.[12]

Too often, philosophers have abstracted our needs, dependencies, and other messy or inconvenient details out of their conceptions of human nature. Reason, will, and other excellences, such as language, have been isolated and privileged as near otherworldly "marks of men." They have been used to signal our separation from nature and to reinforce that humans are not quite part of this world (see "Animals" and "Nature").

Resistance to human-animal continuity

Midgley argues that a much-needed corrective to such abstractionist conceptions of humanness may lie in human-animal comparison.[13] This area of study—which, of course, requires acknowledgment of our kinship with animals—can, properly done, heighten our self-understanding by lighting up the similarities and differences we have with other animals. "Our difference from our relatives," she writes, "may be striking, but the comparison has always been, and must be, crucial to our view of ourselves. It will matter if, as I believe, the gap [between humans and other animals] comes in a slightly different place from where tradition puts it, as well as being rather narrower."[14] As such, our continuity with nature—and also our uniqueness—can more fully be realized not by wholesale rejection of animals as "other," but through careful study and evaluation of our place among them.

Yet, there are numerous challenges in such a venture.[15] Notions of animality tend to display how we distinguish ourselves from other animals. The very term "animal" is value-laden and loaded with symbolism. It has, Midgley writes, always been significant "in forming our communal self-image—our notion of the kind of being that we ourselves are. In trying to define ourselves, we contrast ourselves with something outside us. Accordingly, whatever propaganda humans in a particular age want to put about concerning themselves demands and gets corresponding alterations in the typical notions entertained about non-human animals."[16] These layers of biases, however favorable or unfavorable, make it difficult to see clearly through the various values, concepts, and symbols involved.

Often, though not always, resistance to human-animal continuity is even built into the very terms of the conversation itself. Certain traditions, especially those with more dualistic conceptions of human nature—historically including much of Christianity and Cartesianism—have had trouble accepting our continuity with nature (see "Nature").[17] Though some groups tend toward more continuous understandings of humans and nature—for example, Algonkian, Franciscan, and certain Daoist traditions—Midgley questions those that insistently reject other animals as our kin:

> People do feel a threat here: what is it? Human achievements are secure; they are unparalleled, they are all around us; there is no doubt about them. They cannot be shaken by an inquiry into their roots, elements, and origins. Yet there undoubtedly is a sense of outrage, varying from the disgust of one who finds a caterpillar in his salad to the terror of one who finds King Kong in his back yard, in people's insistence on a firm, simple differentia—a sense that is often felt to make all argument unnecessary.[18]

There are likely many reasons, she argues, that people wish to disassociate themselves from other animals (see "Animals"). One particularly strong reason is that we fear the "beast within" because it represents the things that we most want to push down and out of our lives.[19] We do not wish to acknowledge the rage, ferocity, greed, lust, and deceit that lurk beneath. Fear, guilt, and shame associated with our animal nature lead us to project our undesirable qualities onto other animals, scapegoating them through literature and other devices as something wholly other, something that we are not: ferocity belongs to wolves and lions, not humans, or, at least, not to *me*. Animals are "out there," howling, killing, and dying while humans, safe within the confines of culture, transcend their bestial natures and fates.

On a practical level, Midgley argues that a general lack of experience or contact with animals is likely to blame for these unrealistic, one-sided representations of animals. City dwellers, in particular, have grown less and less familiar with actual animals—real wolves, lions, kestrels, skunks—and know less and less of their natures, needs, habits, and so on. It is easy to abstract generic and idealized groups—birds, animals, insects—over real animals in the world. Philosophers are also to blame here as these symbolic animals are much easier to abstract into philosophical systems that are otherwise incapable of accommodating the complexity of real animals. "*It is the folk-figure*," she writes, "*that has been popular with philosophers. They have usually taken over the popular notion of lawless cruelty which underlies such terms as 'brutal,' 'bestial,' 'beastly,' 'animal desires,' and so on, and have used it, uncriticized, as a contrast to illuminate the nature of man.*"[20] And why would we want to be associated with these beasts, given how they are portrayed? Animals, in this perspective, are symbols of the very things that hold us back from our highest aspirations, be they religious, intellectual, social, or otherwise.

There are, no doubt, many other reasons that humans distance themselves from other animals, including the psychological and practical distancing required to exploit animals—for instance, to eat them—as well as general callousness and indifference toward the nonhuman world (see "Animals" and "Nature"). All in all, however, Midgley argues that the notion of human-animal continuity, which was so forcefully presented by Charles Darwin, is regarded by certain groups as a threat to human dignity:

> Darwin's real offence was to the dignity of *man*. He explicitly wanted to break down the fence that shut off our own species from other creatures. Being a true natural historian, he was deeply aware of kinship with the life around him. He thought this division was arbitrary and misleading. Most of his contemporaries, however, still saw that division as an indispensable defence

against chaos. Though the geologists had succeeded in bringing the Earth within the confines of science, they had by no means tamed its symbolism to the point where it could cease to be frightening. People might now be willing to speculate about the vast and distant processes that formed its history. But it was quite a different matter to be asked to acknowledge kinship with its non-human inhabitants. Those inhabitants—wolf, rat, worm, wasp, raven, serpent, ape—seemed to them chiefly embodiments of the vices. Here, once more, the Earth seemed to link them horribly with human sin. Once more, they shied away from it in alarm. That symbolism, with its attendant horror, still persists today.[21]

Darwin, in other words, touched a nerve. Human dignity here is conceived in relation to what humans are *not*, or at least do not want to be, in the vices, sins, and faults symbolized by other animals. Animals in this sense are the inhuman or the antihuman, the stark contrast to our superior human nature and qualities. Though animals have often been potent symbols of positive qualities (courageous lions and wise owls), negative and distancing symbolism has often had the most force in our visions of the world. We imagine a distance between us and the indignity of other animals, along with a whole range of things that trouble or simply disgust us—not just the conceptual and the symbolic but also the phenomenal and sensual: the sounds, smells, and sights of the animal world; the drives and the behaviors, the passions and the chaos. How could we be part of such a nasty and messy place? Midgley reminds us that, though there are important distinctions and qualifications, we are actually part of this world, however much we resist it. And, if our conceptions of the human condition are to have any relevance in a post-Darwinian world, we would do well to take our animal nature seriously.

Humans among animals

Yet, while it is important to understand and emphasize our continuity with nature as a corrective to unrealistic conceptions of human nature, it is also important to stress that we are not *identical* with other animals. Midgley notes that we are quite unique in our measure of self-awareness, free will, and the extent to which we can reflect on our conflicting motives and desires, among other things.[22] Her refrain is that we share most, if not all, of our characteristics with our fellow animals in some measure—they come from the same source, so to speak. Structural properties, for humans and other species, are multiple and arrange in matters of degree on various spectra. These properties "do not have to be exclusive or necessarily excellent. Nor do they have to be black-or-white, yes-or-no matters. And certainly no one of them is enough alone to define or explain a species. We commonly employ a cluster of them, whose arrangement as *more* or *less essential* can be altered from time to time for many reasons."[23]

Given the diversity of species on the planet, there are virtually limitless shapes and arrangement of these clusters, some more overlapping than others. In the case of humans, we have a remarkable collection of properties, some of which locate us in close proximity with certain animals and more distantly from others, including shared elements of "dolphinity, beaverishness, and wolfhood."[24] In this dynamic understanding of animality in context, the characteristics that have traditionally been trumpeted as our distinctive excellences—thought, language, tool use, culture, self-consciousness, and the like—have real significance but are firmly located within the collection of properties that makes any species (see "Animals").[25] Thus, we should not be astonished when we learn that whales often "improvise" when singing, octopuses can use tools, ants have complex social structures, and crows display elements of self-awareness.

Midgley's approach to human nature is distinctive and significant, in part, because it takes this wider comparative context among animals seriously. She writes:

> *Understanding* is *relating*; it is fitting things into a context. Nothing can be understood on its own. Had we known no other animate life-forms than our own, we should have been utterly mysterious to ourselves as a species. And that would have made it immensely harder for us to understand ourselves as individuals too. Anything that puts us in context, that shows us as part of a continuum, an example of a type that varies on intelligible principles, is a great help.[26]

For Midgley, conceptions of human nature can only make sense in this comparative context. Otherwise, too much is assumed or overlooked. Thus, if one asks, "What distinguishes us from the animals?" her reply is that "nearly everything is wrong with this question."[27] It is rather what distinguishes us "among the animals," and there are *fewer* fixed distinctions than is often supposed. We certainly possess an astounding cognitive structure, in which certain capacities seem to enable increasing structural complexity.[28] This cognitive complexity, along with other phenomenal human traits, is no trifling matter in contemplating our respective place in the world. Yet, she reminds us that these unique abilities have shared roots with all other animal traits and qualities in the wellspring of nature.

The integrated self

Midgley suggests that it is in this context, this background of continuity with the natural world, that we can begin to see the self and some of its various aspects—motivation and behavior, freedom, mind and body, reason and emotion—in more grounded and integrated ways. First, take motivation and behavior. Midgley's general point is that motivation and behavior must fit within a particular context or way of life.[29] This applies to a variety of animals, including humans. *Some* instincts in *some* animals are more "closed." These instincts motivate

> behavior patterns fixed genetically in every detail, like the bees' honey dance, some birdsong, and the nest-building pattern of weaver birds. Here the same complicated pattern, correct in every detail, will be produced by creatures that have been carefully reared in isolation from any member of their own species and from any helpful conditioning. Such genetic programming takes the place of intelligence; learning is just maturation.[30]

Of course, no instinct can be completely closed off from environmental conditions, but, for the most part, closed instincts are causes of very specific types of behaviors with little to no variation. Midgley distinguishes these motivations from "open instincts," which are much more predominant in humans. These highly variable instincts are something like "programs with a gap. Parts of the behavior pattern are innately determined, but others are left to be filled in by experience."[31] Open instincts are, as it were, "general tendencies to certain *kinds* of behaviour, such as hunting, tree-climbing, washing, singing or the care of the young."[32] Experience, imprinting, and learning *complete* the range of possibilities of an open instinct, highlighting how innate and environmental factors are much more closely entangled than some have assumed. This especially is the case with animals with rich social lives, such as humans.[33]

Thus Midgley argues that we need not see innate and environmental influences on human behavior as mutually exclusive. This is akin to insisting "that the quality of food is determined *either* by what it is like when you buy it *or* by how you cook it, but not both."[34] Here, she resists the portrayal of culture as something antithetical to nature. Instead, she insists that culture grows out of our social nature. Humans, in this picture, are "half-finished." "We are naturally culture-building animals," she writes, "but what we build into our cultures has to satisfy our natural pattern of motives."[35] Biological and cultural factors, in this two-way street, are interrelated and interactive rather than competing alternatives, working in concert to shape and mold us in all of our sameness and uniqueness. If we want to understand our various motives (instincts are one type of motive among many, and Midgley focuses more on "motives" than "instincts"), we must contextualize them in the particular circumstances in which they operate "like the interaction of currents in the bed of a river."[36] Explaining motivation in this fluid context thus requires a wider map or a "comprehensive plan of life, just as explaining a single act is placing it on the map of a smaller area—a partial plan, a motive. And the explanation is adequate if it can fit it in without distorting the normal arrangement beyond the bounds of what is credible."[37] Motives, in this way, cannot be reduced—as stemming from purely innate *or* environmental sources—or understood in isolation.

Yet, the temptation in taking motives seriously is to isolate and oversimplify, which has long been prevalent outside of the confines of the nature-nurture debate. For all the useful analysis that theorists such as Nietzsche and Freud may have provided in understanding the strength and influence of motives in human behavior, Midgley argues that they ultimately missed the mark "by *reducing* many motives to one or a few basic ones—sex, self-preservation, power. . . . Human life simply contains more motives, even more separate groups of motives, than they allow for."[38] She reiterates:

> I cannot exaggerate how complicated human nature is. But saying there is such a thing as human nature isn't saying that there is some *vast* motive for everything. We are animals, and animals are in this complicated situation of having a lot of motives. I don't think that they can be boiled down. That's obviously helpful when you've got a conflict between some of these motives—to try and see how and what they share—but it isn't going to be final.[39]

Actions must rather be explained within a wider structure of motivation and as part of the whole character in context.[40] The plurality of motives in human life and behavior, as such, makes situating any given thought, emotion, or action more complicated than reductionist accounts assume. People commit atrocities, create symphonies, and donate to charities for whole sets of reasons that are irreducible to one explanation. Of course, the reasons are dynamic and differentially weighted, but no law decrees that the will to power, rational self-interest, or sexual impulses are primary. Rather than reduce all explanations to a seductively elegant or unifying motive, Midgley argues that we must work harder—and across disciplines—to bring together the many complex strands of human nature into an intelligible whole.

Freedom and determinism

Reduction is one of Midgley's most consistent targets. This is not just because of the one-sidedness of reductionist claims but also because of the reactions these claims provoke. Reductionists of various sorts—who reduce or simplify motivation and behavior to one underlying factor or set of factors—claim that the biological or environmental determinants of human nature may well explain away such things as free will.[41] Here, we are so controlled by genes or society that we have no real choice in who we become or how we act. Freedom is, as such, an illusion. This, quite naturally, leads what Midgley labels "obscurantists"—those who resist or deny naturalistic explanations of humanness—to react against this position and radically *elevate* human freedom. Thus the extreme conclusion is that there really is no such thing as "human nature." We essentially choose or will what we become, regardless of other internal and external influences.[42] Midgley, again, rejects each extreme. She argues that reductionist accounts of human freedom, on the one hand, are often manifestations of *fatalism*, the ideology that all events are inevitable and that humans are impotent in the face of this reality.[43] She views determinism in a much softer sense, saying that it is "simply the modest assumption of that degree of regularity in nature which is necessary for science."[44] She opts for the compatibility of free will and determinism, arguing that fatalism, rather than determinism, is the only real threat to freedom:

> Determinism is of course often taken to be incompatible with a belief in free-will. But it seems possible that this idea springs from a superstitious over-inflation of both concepts, particularly of determinism. A melodramatic tendency to personify physical forces and other scientific entities can represent them as demons driving us, rather than humble general facts about the world, which is all they have a right to be seen as. This produces fatalism, which certainly is incompatible with a belief in free-will, since it teaches that we are helpless in the hands of these superhuman beings.[45]

Freedom in her account is not tantamount to radical unpredictability but rather, more modestly, involves understanding action within the context of a range of "facts about the world," which include numerous causal factors—genetic, social, economic, and so on.[46] None of these causal factors, in the end, has complete or overpowering influence in human life.

Against obscurantist accounts, on the other hand, Midgley argues that human nature is *necessary for*, rather than *opposed to*, free will. "To be free," she argues, "you have to have an original constitution. Freedom is the chance to develop *what you have it in you to be*— your talents, your capacities, your natural feelings. If you had no such particular potentialities in you for a start, you could have no use for freedom and it could not concern you."[47] Our personalities, in turn,

develop as unique individuations of the general influence of internal
and external factors.[48] Thus, we are not fatalistically determined, as
reductionism suggests. Nor are we radically indeterminate, as some
libertarian notions of freedom would suggest (sane choice, after
all, seems to be, at a minimum, connected to the "determinant"
of our personalities). Rather, we are able to make meaningful, free
choices—or "efforts"—in the midst of many partial factors:

> [Free will] is not just a name for a general exemption from causality.
> Essentially it concerns *effort*, which is a perfectly real causal factor.
> It does not call on us to claim that outside causes of action don't
> exist or that they have no effect on thought. It simply means that
> they should be treated as the partial factors that they are, not as
> all-powerful tsunamis that must overwhelm mental strivings and
> intentions. It means that our efforts can in principle be effective;
> that thoughts have their real place among other kinds of causes
> in the world.[49]

Our freedom, ultimately, exists in a context that both gives rise to and,
to some extent, limits it. What results is something of a moderate
freedom that is contingent upon the many influences on our behavior.

Mind and body

Now to shift away from motivation and behavior, which I'll return to in
"Morality," to look at some other aspects of the integrated self, which
include the continuity of mind and body and reason and emotion.
Midgley does not presume to solve the mind-body problem—if, indeed,
it is even something to be "solved"—but rather to avoid the more
fragmenting conclusions that arise within its dualistic framework. She
essentially argues that the Cartesian legacy has unnecessarily skewed
the debate toward unrealistic and distracting extremes. Descartes,
who separated mind and matter as two distinct substances, could
not explain or conceive of the conscious mind in the materialist
terms of seventeenth-century mechanistic science.[50] Mind, as
wholly separate from physical matter, essentially removes humans,
as "thinking things," from the rest of nature. Thus, when we talk

about consciousness or mental properties, we are talking about the characteristics of the substance of mind rather than matter. Bodies, on the other hand, classify as matter and thus are of a different substance altogether. Talk of minds, since Descartes, has consequently been dominated by the notion of two competing substances: mind *or* matter. This, she argues, severely limits how we think of the mind. In this antithesis, we are forced to choose between stark alternatives. The mind is either purely material—something accountable to the systematic study of science, and which may ultimately be a chimera[51]—or something more mysterious, a separate substance, akin to the soul, with perhaps no intelligible link to the body. Thus, the more scientifically oriented option must be to reject the substance of the mind completely in favor of a thoroughgoing materialism.[52]

Midgley argues that we should resist the assumptions of substance dualism if we are to think of the self in more integrated and coherent ways. Dualist positions, in accentuating the nonmaterial nature of the mind (or soul), overemphasize the subjective or "inner" elements of selfhood over the objective or "outer" elements of the physical brain and body. Materialist positions, on the other hand, reject the substance of mind and conceive of brains and bodies as objects alone—the substance of matter—often discounting the inner or subjective aspects. Midgley argues that each selects and emphasizes one *aspect* of our life and experience. She thinks it better to reorient our discussion of "mind" altogether. Mind, she argues, might better be conceived of as a way of talking about the subjective aspects of the self. She writes:

> I think it has become clear that both these solutions [dualism and materialism] are equally unworkable. We have to avoid dividing ourselves up as Descartes did in the first place. *Things go wrong as soon as we start thinking about mind and body as if they were both objects*—that is, separate things in the world. The words mind and body do not name two separate kinds of stuff, nor two forms of a single stuff. The word *mind* is there to indicate something quite different—namely, ourselves as subjects, beings who *mind* about things. The two words name points of view—the inner and the outer. And these are aspects of the whole person, who is the unit mainly to be considered.[53]

In this way, we can conceive of mindedness in reference to the subjective, inner aspects of personhood that come together in the *whole person*. Mind and body are not, in this sense, isolable as separate substances, but aspects of this whole. Midgley argues that the dilemmas of mind *or* matter and dualism *or* materialism make it seem as if there is a final choice that must be made between these aspects. The self, in this way, cannot help but be fragmented. Consequently, Midgley maintains that both approaches, as carryovers of the Cartesian obsession with consciousness, lose track of the self or the whole embodied person in which mental events occur.

Reason and emotion

Midgley likewise positions reason, an aspect of this minded self, within the overall structure of our lives. Reason is a *significant* part of the self and plays a vital role in our lives, including the moral process (see "Morality"). Yet, reason and other aspects of the "intellect" or "thought" must be realistically balanced with emotions, feelings, and other aspects of what we might call "heart." The practice for some in the Western intellectual tradition has been to demarcate and compartmentalize reason and emotion as completely separate and antithetical to each other. Champions of reason such as Plato, Descartes, and Kant glorify its role in ruling the self, as something of a colonial governor in the foreign land of passions.[54] Alternative accounts, which favor the primacy of emotions, then react against this colonial model, demoting or downplaying the centrality of reason in relation to the emotions or passions. Hume, for instance, said that reason was the "slave to the passions" in the moral process. The reason *or* emotion dichotomy, posed in this way, can make it seem as if we must choose which will be the hero in the great stage drama of human life.

Midgley rejects this dramatic opposition and proffers that reason is rather the "containing scene of the drama" or the "whole well-ordered gathering" of the self in which the various passions are related and prioritized.[55] In this scenario, reason is more of an arbitrator or facilitator in integrating the self, akin to the reflective elements in Darwin's account of morality (see "Morality"). Here, reason does not simply await the commands of these passions nor, conversely, rule the passions but rather arbitrates between and integrates them into action. Thought, feeling, and action, in this way, are aspects of an integrated personality, not warring tribes.[56]

Midgley argues further that reason and emotion cannot be sharply separated—they are intertwined and each shares elements of the other in an interrelated structure.[57] "All thought," she writes, "involves feeling (for instance, feelings of satisfaction or dissatisfaction with an argument). And nearly all feeling involves thought (for instance, anxiety nearly always includes the thought of some specific disturbing circumstances and possibilities)."[58] Since reason and emotion are so entangled and interdependent, neither can be rigidly separated and isolated as *the* primary driving force in human life and action. In this way, we do not have to choose whether we are going to be people that are *ruled* by one or the other aspects of our nature. It is rather a matter of *emphasis* whether we focus on the intellectual or emotional aspects of our selves. Otherwise, "if the relation between these two is seen as unintelligible, then there is a blank gap within each of us, a chasm across which negotiations can only take the form of conflict."[59]

The relational self

These elements of human nature and the self—motivation, behavior, freedom, mind, body, emotions—are by no means exhaustive of Midgley's far-reaching account of the human condition (see "Gender"). They are, however, a start, and set up a number of issues and topics (morality, etc.) that remain to be covered in this book. The remainder of *this* chapter is devoted to another significant aspect of the self: relationality.

Critique of individualism

Midgley's emphasis on the relational aspects of the self emerges from her critique of (hyper) individualism. In social and political thought, there is a long-standing impulse to hyperbolize the centrality of either the individual or the collective, the self or society. Those that stress the individual tend to favor accounts of personal freedom while those that stress society tend to favor accounts of collective solidarity. Hobbes, Nietzsche, and Sartre favored the former—the "solitary" or "heroic self"—while Marx favored the latter. Each of these positions, in keeping with the prevailing patterns in the warfare of ideas, is prone to over-dramatization and exaggeration. While Midgley warns of the dangers of hyper-collectivism—of subsuming the individual into the collective—she is especially concerned with the implications and damages of exaggerated individualism, which unrealistically isolates individuals and thus fragments our understandings of the self.

Two broad types of individualism are often entangled and interrelated. The first type of individualism is "factual individualism," or psychological egoism, which is a descriptive account of human sociality: humans *are* individualistic.[60] The second type of individualism is "moral individualism," or moral egoism, which is an ethical stance on human sociality: humans *should be* individualistic. Midgley is critical of numerous forms of individualism—factual and moral, ideological and political—because they often rely on each other. Her main targets, though, are typically those rooted in atomistic and contract thinking. Social atomism, in brief, is the idea that people are distinct individuals. Like atoms or billiard balls, people are seen as discrete, self-contained units. Society, in this sense, is nothing more than the aggregate of individual units. In its more egoistic forms, self-interest is portrayed as a universal law of nature and one of—if not *the most*—central features of human nature.

Hobbes's theory of human nature is one of Midgley's most regular targets. This is because it is so vivid, influential, and, in her estimation, flawed. Hobbes claimed that, in the state of nature, we are egoistic and largely motivated by self-preservation and survival. This underlies his formulation of the social contract, in which he argued that it is in our best interest to strike a collective bargain in order to escape this ruthless state. Otherwise, there is little stopping us, in the warfare of all against all, from pursuing our interests to the destruction of others. Life, if our egoist natures were to prevail, would be "solitary, poor, nasty, brutish, and short."[61] Thus, we must contract with each other to *live* in a lawful society (under the sovereign). We then give up some of our more dangerous freedoms—for instance, the freedom to kill and steal—to live in a lawful world that is, in the end, in the interest of our self-preservation.

Other prominent variations of the social contract—most notably John Locke's and Jean-Jacques Rousseau's—also emphasized the innate and original individualism of the human condition.[62] This and the general rise of contractarianism make sense historically. The violations of church and state over personal liberty instigated calls for freedom and recognition that each person (or each man, at least—see "Gender") was a unique individual and created equally. These enduring principles gave rise to political and legal reforms that protect liberty and human dignity to this day. In this way, the political benefit to the individual, in the sense of freedom from certain abuses of the state, has proven invaluable.

Midgley does not question or disregard these advances in freedom and dignity. What she does question, however, is the desirability of a type of individualism that has gone well beyond the original purposes of the concept. She argues that an emphasis on freedom and the social contract is not necessarily conducive to the types of lives that people often want to live. She writes:

> In personal life this is not half so helpful. The trouble there is that not all our relations with the people round us are power relations. Although we do very often want to be free from their demands, we also badly want to be free to make demands on them, and all these demands form parts of patterns that, as a whole, we may still want to be involved in. If you play the violin you need orchestras; if you like to argue, you need an opponent. However tiresome other people are, we do not really want to get rid of them. When we worry about this, it surely emerges that freedom, as an ideal, is merely a blank form, negative and neutral, a name for getting rid of something or other. Its meaning is only clear when we specify just what we want to be free *from* and free *for*.[63]

Midgley's point is that the quest to be free *from* things—interference, harm, obligations—can obscure what we want to be free *for*: meaningful relationships, care, trust, support, and so on. We then run the risk of cutting ourselves off from many of the things that contribute to our sense of self and meaning in the world. Contract thought arose because of real threats to freedom in the world, but, she argues, its political leverage should not be confused as a policy covering all our social relations:

> Plainly, social contract thinking is no sort of adequate guide for constructing the whole social and political system. It really is a vital means of protection against certain sorts of oppression, an essential defence against tyranny. But it must not be taken for granted and forgotten, as a safe basis for all sorts of institutions. It needs always to be seen as something partial and provisional, an image that may cause trouble and have to be altered. It is a tool to be used, not a final decree of fate or an idol to be worshipped. It is, in fact, just one useful analogy among many. It must always be balanced against others which bring out other aspects of the complex truth.[64]

Relationality and interdependence

Midgley argues that atomistic and contractual visions misconstrue our social landscape not only by harboring hyperbolic notions of freedom and individualism but also of competition. That we are, first and foremost, individuals fighting for limited resources, whatever they may be. In this vision, we are encouraged to see each other—and other animals and the natural world, for that matter—as competitors rather than fellow beings. We engage each other through self-interested and easily dissolved contracts. Relationships are thus provisional and instrumental, things we do not necessarily seek out and need, but things we use. Seeing each other in this contractual manner overshadows social features that entail more than calculated self-interest, such as cooperation, solidarity, altruism, responsibility, and mutual care.

Midgley argues, furthermore, that egoism cannot be the overriding feature of human nature that Hobbes and others assumed it to be. We do have egoistic and competitive *aspects* of our natures, no doubt, but this is not the *whole* picture. To even exist as a social species, we require much besides ruthless competition and selfishness. She points out that the word "selfish" is itself a negative term, indicating a lack of something, such as lack of regard for others or lack of cooperation.[65] This makes sense as we are bond-forming creatures that group together and need each other at every stage of life. "Far from being originally solitary," she writes, "the earliest human beings were heirs to a long, complex tradition of group life, deep social affection and interdependence, a tradition which dates from many ages before their emergence as a separate species and their famous rise in intelligence."[66] We have many needs, some of which, such as freedom and solitude, are more self-contained. But many or most of our deep and overarching needs require other people—bonds, love, attention, companionship, sex, and so on. Without these, the very self that individualism so highly values risks being fragmented and disintegrated.

Midgley's alternative vision to hyper-individualism is a relational approach that balances the interdependence of the self and others. The relational aspect emphasizes the variety of intrinsic relationships one has with others—with family, friends, communities, and the nonhuman world. The interdependent aspect emphasizes the webs of dependencies and interconnections that exist within this network of relationships. These networks of relationships and dependencies mold and shape us throughout our lives. Yet, this network is dynamic, as we also shape others and the world around us. However active or passive we are in this process—for instance, we are almost completely dependent on others as young children—we are likely never as independent or isolated as atomistic and contract thought would lead us to believe.

This need for others does put some limitations on human freedom and pushes back against the most radical and libertarian notions of free will. But, as rational agents, we do still choose and forge our paths in life and are able to exert an influence on the world. Midgley's point is that these paths are never in isolation or without a wider map or landscape. They take place in a social context that is largely constituted by an ever-changing and meaningful landscape of relationships and dependencies. The isolated ego myth overlooks or ignores this landscape in favor of a simple map of atomistic agents. Much like LEGO blocks, people are here imagined as fitting together and separating with ease, forming and absolving contracts to suit our needs. Midgley resists this plastic vision, arguing that we cannot snap so easily in and out of each other's lives. We are instead much more deeply connected as "members one of another."[67] With this context in mind, she writes, "It may even become possible for our species to admit that it is not really a supernatural variety of Lego, but some kind of an animal. This ought to make it easier to admit also that we are not self-contained and self-sufficient, either as a species or as individuals, but live naturally in deep mutual dependence."[68]

The many connections and entanglements that we acquire throughout our lives exist not just between the self and other individuals, but also between the self and various collectives. Rather than envisioning ourselves as soloists, we can see ourselves as parts of a larger dance in which our own distinctive moves and rhythms intertwine with those of others and the whole. "Our life," she writes, "is not a collection of solo performances but an immensely intricate large-scale dance in which solos take their place among figures performed by groups of the most varying sizes."[69] Our places and roles in this dance may shift and change, but our unique expressions as soloists are cultivated *within* the larger dance itself. The dance and the dancer need and complete each other. Ideally, neither overrides nor overpowers the other. Rather, they complement each other in the fullest harmony. This dance is embedded in a larger whole still, composed of many other entities and collectives, from our local communities to the global ecological networks of which we are part. If we abstract individuals away from this setting, we again lack the total picture through which we can envision the wholeness of the self:

> This "whole person" of whom we have been talking is not, then, a solitary, self-sufficient unit. It belongs essentially within a larger whole, indeed within an interlocking pattern formed by a great range of such wholes. These wider systems are not an alien interference with its identity. They are its home, its native climate, the soil from which it grows, the atmosphere which it needs in order to breathe. Their unimaginable richness is what makes up the meaning of our lives. *The self's wholeness is not, then, the wholeness of a billiard-ball but that of an organism*, a transient, struggling creature which has, of course, its own distinct shape but which still belongs in its own context and background.[70]

In this context, though we value our freedom and individual pursuits, they are embedded within these inescapable connections to the whole. This does not mean that our identity is equivalent or subsumed to the collective. This hyper-holism is as unrealistic to Midgley as hyper-individualism. Rather, the search for wholeness involves negotiating the interplay of collectives and individuals, between "compressing individuals into a homogenous mass and isolating them completely; between lumping and splitting."[71] In other words, as an antidote to hyper-individualism, a more integrated understanding of the self involves mapping its many interrelations and interdependencies without alternatively over-exaggerating the collective. The self, as such, can only make sense in this comprehensive context, and the tempting extremes of both individualism and collectivism flatten out the complexity in between.

The wholeness of the self

This type of resistance to polarizing notions that divide aspects of the self is a hallmark of Midgley's thought. The antitheses in this chapter—biology versus society, human versus animal, freedom versus determinism, reason versus emotion, mind versus matter, self versus others—are prominent among others that remain to be covered in subsequent chapters: human versus nature, male versus female, science versus religion, and so on. Rather than trying to "unscrew the outside from the inside of the teapot," Midgley's project seeks to connect the various parts of our selves—thoughts, feelings, actions, relations, and so on.[72] Dramatic exaggerations of aspects or elements of our nature are always tempting, as they typically contain real insights and partial truths into our nature. Yet, when these dramas have arisen, Midgley was usually there, in her characteristically balanced and sober way, to reel things in, as it were. This is not just because we need to understand ourselves in more coherent and integrative ways but also because "the way in which we think of ourselves—the picture we form of our essential nature—directly affects the way we live."[73] This entails morality, which I explore in the following chapters.

3

Morality and Wholeness

Human life is terribly full of contraries. They are all around all the time. You resolve one and a couple more crop up.[1]

Midgley's approach to morality is interconnected with her conception of human nature and the relational self. Morality, in this framework, can be broadly described as that which deals with what is *serious* or *important*, and moral philosophy is the process of working through central issues in our lives. Drawing on Darwin, she positions morality in our deeply social nature. It is here that we seek to reconcile some of our many conflicts in pursuit of a more integrated life. However, this reflective process is far from simple or straightforward. Quite often, the difficult work of weighing and relating values and claims is oversimplified in an attempt to manage the complexity and occasional chaos of life. In the process, important elements of life and morality can be neglected and overshadowed by the ideals of individualism and freedom. Midgley resists this fragmenting approach to morality much as she did in relation to human nature and the self. She tackles these and other simplifications— such as moral skepticism—and argues that moral philosophy must involve the whole person in context. Understanding these dynamics may help us, in the end, to navigate the possibility of moral change in society. Cultures do have the resources to flourish, though these values and ideals often require thorough scrutiny and reexamination.

Moral issues and philosophy

Midgley conceptualizes morality as involving what is serious or important. Thus, to call something a moral issue highlights the special attention that should be paid to it.[2] Serious matters affect us deeply and connect to our character or "central system of purposes."[3] In this picture, morality is intimately connected to our nature, involving what is important for us in relation to our aims and needs in life. This does not mean that moral issues are always clear and fixed. Many things are important to us, and once one begins to consider their relative importance, it becomes apparent that there is no "hard and fast line fixed round the moral. Seriousness is a matter of degree."[4] Yet, clearly, not every issue is a moral one. Lest this conception of morality allow for too much, Midgley reiterates that morality is public, and the seriousness of issues in human life must be publicly defensible: "Morality cannot be private whatever else can, and if anybody claims that some unexpected thing is either *very serious* or a *moral question* to him, he has got to explain what his system of purposes is and how this comes into it, otherwise we shall not understand him."[5] Thus, if an issue can be shown to be serious within an appropriate system of purposes, it can reasonably be seen as a moral issue.

Large issues such as war, climate change, and poverty certainly bubble up as serious matters, but some seemingly innocuous events, such as moving houses, may also classify as important. If one moves down the street, it may not be such a big deal. However, if one moves houses from an urban to a rural area, or from an affluent neighborhood to a low-income one, or from a diverse neighborhood to a more homogeneous one, the move is much less innocuous and connects to whole sets of purposes and ideas about how to live.[6] These moves involve choices regarding lifestyle, status, community, and other central features of our daily lives. Likewise with certain dietary choices, where we choose to send our children to school, what profession we choose, and how we vote.

In this way, all manner of moral decisions and actions are intertwined in how we navigate the world, both at grand junctures—say, Brexit or the 2016 US presidential election—and in our everyday lives—should I eat meat today? The "moral point of view" is, in turn, the perspective that takes into account the relative priority of matters as more or less central to life, relating them to the whole person.[7] Or, in more Aristotelian terms, the whole character. Morality, in this sense, is not a separate realm or special content per se—happiness, for instance—but involves and relates content as *more or less* serious or important: What matters and why? How does this particular issue fit into the larger whole of our priority systems? What do we do about it? For Midgley, moral thinking is, accordingly, a connective and integrative process in which we weigh and evaluate goods and harms in relation to the whole of life. It is through this process that we can stand back and wonder, "Would it perhaps be a bit better different?"[8] Our approach and answers to this question place us on the path to moral philosophy.

Moral philosophy, among other things, involves making and following ethical arguments, stating and generalizing standards, and connecting those standards to life as a whole (including our wider systems or forms of thought).[9] Though much of modern moral philosophy is preoccupied with the first two stages of argumentation and generalized standards, Midgley stresses that the third stage— connecting these standards to life as a whole—truly "completes" the process and is perhaps the most challenging and neglected part. It requires zooming out from particular moral issues and principles and placing things on a wider map. Once issues are positioned in this more comprehensive way, we can give reasons as to why the world should be one way rather than another, why *this* world would be better than *that* world. "Here argument is called for, and here moral philosophy starts."[10]

Arguing for a particular type of world—for a particular vision—is easier said than done. It seems that as much divides us as unites us, and various land mines abound in moral thinking: Should we be guided by reason or emotion? Are we inherently selfish or altruistic? Do we value freedom or bonds? What do we make of the values of other cultures? Midgley reminds us that we are capable of working through the many tangled concepts surrounding moral issues, drawing out their confusions, and moving toward workable solutions.[11] Yet, we will likely not find tidy resolutions to many issues because life itself is not very tidy.

Morality and human nature

Our answers to moral questions may be incomplete and conflicting, more provisional than final. Yet, these resolutions, however tentative, are important if we want an integrated life. Morality is, in large part, the business of sorting out what matters. If we wish to ask questions regarding how the world should be, it makes sense to keep in mind considerations relating to the type of beings we are—our needs, purposes, possibilities, and limitations—and what is important to us. This does not mean that morality is a matter of simply looking at the facts of our nature and, in turn, knowing how to act. As critics of naturalistic approaches to ethics have long argued, simple or crude "evolutionary ethics" often conflate natural facts and values in an illicit way that distorts moral reasoning. What *is* evolved is not necessarily the same thing as what *should* be. (Social Darwinists argued that we *are* and *should* be selfish and egoistic—critics have challenged both the "is" and the "ought" of this claim.)[12] Rather, Midgley makes the more modest point that understanding our nature, insofar as this is possible, is central in how we think about our place and possibilities in the world. Indeed, this is how we can even make sense of what is and what is not *important*. Of course, there are good reasons to be skeptical about appeals to human nature, especially since these appeals have historically often reflected an unjust status quo (slavery, sexism, etc.). Human nature can be distorted and abused in support of many objectionable ideas and policies. Midgley echoes these concerns but does not conclude that the concept of human nature must be abandoned altogether. Rather, her position is that it can be approached in more complex and nuanced ways, salvaging its important role in assessing and critiquing our actions and institutions.[13]

Darwin and morality

Midgley, accordingly, turns to Darwin for help in drawing out some relevant insights on our nature.[14] Darwin was deeply immersed in many aspects of human evolution, including its implications on moral development. His theory of morality, which integrated ideas from

Aristotle, David Hume, and Immanuel Kant, among others, offers a compelling approach to moral development, which stems from our social and reflective natures. His account, moreover, connects morality to our efforts to integrate our lives, a reflective process that is virtually inevitable in rational beings such as ourselves. Midgley thinks this account is worth exploring not only because of its insights into our nature but also because of how it counters exaggerated portrayals of "Darwinian morality." Midgley resists both reductionist and obscurantist approaches to Darwinian morality, each of which distorts his position and contributes to the general confusion regarding morality and evolution. Rather than exaggerating egoism and competition alone—as reductionists would have it— Darwin emphasizes our social sentiments.[15] These social feelings and sentiments (empathy, sympathy, compassion, etc.) supply something of the "raw material" for morality.[16] As such, they are not morality proper, but are further channeled and shaped by reflection (and language), providing motivation and, to a certain extent, direction. An emphasis on competitive egoism, Midgley argues, is more representative of Hobbesian individualism than Darwinian morality (see "Human Nature"). Against obscurantist positions, which argue that morality and human behavior are discontinuous with nature— that the origins of morality cannot be explained in naturalistic terms— she argues that Darwin provides a sensible account of how morality could indeed arise in an evolutionary context. As such, his account appreciates and retains the complexity of morality while firmly locating it in a natural context.[17]

Darwin locates the roots of morality in the drive to reconcile the many conflicts within our complex motivational structure. Human motivation is not only complex but also plural. There is no overriding motive such as self-preservation, sex, or the will to power (see "Human Nature"). As such, we need some way of organizing and prioritizing these many motives, which can and do conflict. Motives and desires, which reflect our many needs and aims, vary in intensity, duration, and centrality. In certain cases, intense and short-lived motives—such as panic or fear—can conflict with more central social motives. We can consider care for young. In some species of birds—Darwin mentions swallows—the impulse to migrate may be so intense and central that they ultimately abandon their young and follow the call. Other birds, such as shoebills, may favor one chick over the other, which is only kept around as a backup plan, so to speak. Once it is apparent that the first chick has the best chance of survival, the other is abandoned or otherwise neglected. These birds, in this way, are not immoral, but amoral. They lack the proto-moral ability to prioritize and reflect on their conflicting motives.

Humans, on the other hand, are far less likely to abandon their young, whatever the reason. We are able, within the bounds of circumstances, to make choices between impulses, to decide how to act. Few impulses can ever be strong enough to lead us to abandon or otherwise harm our children. And, when this does happen—as it tragically can and does—we strongly condemn it. This is not just because we have powerful instincts to care for our young (much like birds do), but because we can *reflect* on our various and often conflicting motives and, within reason, *prioritize* certain ones over others. Self-preservation may compel us to flee from something (a fire, predator, or other danger), but we are likely to save, fight, or otherwise tend to our children—and often others—sometimes at any risk. Though we are variously moved by intense and fleeting short-term motives, the enduring and central *social* motives, which have deep roots in our evolutionary past and have resulted in profoundly affective bonds and sentiments, have more lasting effects on our lives.[18] "They are a deeper," Midgley argues, "more integral part of our nature. They are something more central to our characters than the passing impulses that often overwhelm them."[19]

In this context, Darwin sees the rise of the moral sense or conscience as inevitable, given our social *and* reflective (rational) nature.[20] Abandoning our young or even a stranger to their fate leads to inner conflict and feelings of disintegration. The process of grappling with and reflecting on the inner conflicts that arise from this dissonance contributes to the development of moral thinking, which, over time, is further elaborated, refined, and systematized. We reflect on conflicting motives and articulate rules and duties to guide our actions (we ought not abandon our young, we ought not callously overlook others in need, etc.). Midgley writes: "What put 'the imperious *ought*' in the human vocabulary is not ([Darwin] says) primarily fear of punishment by society or parents, but horror at one's own conflicting and sometimes destructive motives."[21] The golden rule—do to others as you would have them do to you—thus emerges as a response to our prioritization of the social instincts, of treating others well.

There are no "ready-made" solutions for dealing with this striving to satisfy "our whole being."[22] What is needed is some system of priorities that can help sort and evaluate our many motives and provide some type of guidance on how to act on them. We develop ways of understanding moral choices, even if, at times, we may only have the rough guidance of "better or worse" in the haze of ambivalence and complexity.[23] This process involves working with others and relies on our "willingness and capacity to look for shared solutions to them." In this way, "morality, as much as language, seems to be something that could only occur among naturally sociable beings."[24] Cultures, however imperfectly, thus develop "sets of customs" that elaborate priority systems and help us settle clashes and facilitate conflicts.[25] "It is," Midgley argues, "because there is no pre-set, universal priority system available that cultures differ so much." Yet, "it is because their basic problems are still the same that cultures are, none the less, so similar."[26]

Value systems and moral claims

Understanding the development of morality in this Darwinian manner does not provide us the answers we need to make moral choices (though it does illuminate how the development of morality is deeply connected to our social natures). We must still, in the end, *choose* better or worse ways of acting and being in the world. And, Midgley remarks, "the business of choosing—what you do in ethics—cannot be explained by saying 'That's how we've evolved.' We've always got a choice to make at the time. And making the choice is an art that we have to cultivate."[27] For these purposes, we need workable priority systems that help us weigh various goods and harms relative to their central importance or seriousness in human life.[28] And what is important to us depends on the type of beings we are. Though the facts of our emotional and rational constitution do not directly lead to our values, they can help us in understanding them. And this understanding, she argues, "*does* give us practical guidance" because "values register needs."[29]

As beings of a particular kind, some things are better for us than others, but they all combine in important ways to form a whole or integrated life.[30] Freedom and meaningful bonds, for instance, both matter in a full human life, along with a whole range of basic needs such as food and shelter. Approaches that prioritize one or two elements in our value systems—dignity, maximizing pleasure, and so on—mislead us by simplifying the many things that are important to us. Priority schemes help us to sort through values and direct our efforts by emphasizing and prioritizing some over others in relation to our aims and purposes.[31] However, *priority* signifies just that, and often the choices we are forced with in everyday life are ones of *better or worse*, as opposed to *the good or the best*. Here, compromise and particulars rule the day over singular solutions and universals. Fortunately, she reminds us, we are not alone in working through these systems. Values are public, and moral judgments are made intelligible by elucidating them within the context of shared social worlds.[32] As such, we have a rich source of solutions, some better than others, for working through practical moral decisions.

Moral claims, in these value systems, are adjudicated in particular circumstances with their attendant context and considerations. Should I eat fish? Should I donate money for a cause? Is my leisure time more important than volunteering in my community? Should I report my colleague's conduct? Should I vote for this or that candidate? Maybe I should bike or walk instead of driving today. Should I purchase something from this company that exploits its employees? Is this immediate gratification more important than my friends, family, work, community, and so on? Here, there is no direct moral calculus that can give us all of the answers all of the time. We must place things on the wider map of our lives and allow for a range of solutions and guidance. Gesturing toward a moderate form of pragmatic pluralism, Midgley addresses the need for seeing the forest of morality through the trees:

> Morality is not a single tree; it is a forest made up of many . . . all of which need to be mapped and understood as a whole. That, indeed, is why I recommended earlier that moral philosophers

should deal in whole maps, rather than in single dividing-lines. And these maps must also allow—just as the many maps-of-the-world at the start of our atlases do—for the confusing fact that this forest, like other forests, changes constantly along with the life around it.[33]

Mapping this terrain is an arduous, long-term process in which we constantly evaluate and incorporate situations and particulars as they come. Along the way, we must take into account a range of claims from a range of beings and entities—human and otherwise—while being realistic about our moral capacities and limitations (see "Animals").

Pluralism and the complexity of moral issues

It is likely clear by now that Midgley does not advocate for any one particular approach to morality, such as utilitarianism, Kantianism, or virtue theory, though she is favorable toward elements of each in varying measures and leans more toward virtue and Aristotelian approaches. A couple of reasons stand out for this resistance to singular moral systems. The first reason involves the challenges of foundationalism. She is skeptical of attempts to ground morality on any single premise or base, as any such premise is unlikely to hold up as an indisputable foundation on which we can build our entire moral framework. Talk of foundations is also misleading and misdirected, as it portrays singular moral premises as a solid bedrock on which we can build or pile upward. While there is value in stable premises, Midgley remarks: "My heart always sinks when that starts because this approach may be an important bit of it, but [morality] is not gravitation. It's not a pile. It's not linear."[34] The image of foundations is simply too limiting here, since, after all, structures tend to have one foundation. If morality is seen as a single structure with a single foundation, this would require demolition of the whole if a new foundation were to be implemented. In this scenario, moral philosophy can become a zero-sum game of perpetual deconstruction and reconstruction.

She prefers organic or cartographic imagery, which allow for layers of complexity instead of monolithic structures. These also invoke more cooperative and exploratory visions in which various aspects of moral thinking can be appreciated and, when possible, brought together.

In addition to these objections to foundationalist accounts of morality, Midgley argues that no single system can universally and unilaterally account for the range, immensity, and complexity of moral issues. Influential moral systems—contractarianism, rights theories, and so on—arise and are valuable as ways of dealing with particular problems. Yet, they can be confused as covering the *whole* of morality. This makes sense. When moral philosophers discover some particularly important insight that has potential to help solve a range of issues, they naturally wish to expand and universalize its usefulness for others. "When confronted with this highly muddled world," she remarks, "we try to find balance, and if we find one, we try to fit other things in with it and make it all tidy."[35] In discovering important insights and concepts, we might wish to "spread the system, which gets more dramatic and proliferates."[36] This endeavor, as such, can yield some important insights and supply some valuable conceptual tools in dealing with a wide range of issues.[37] Yet, for all these insights and applications, single answers or sets of principles can distract us, again, from the whole. The most eloquent advocates of this type of reduction—perhaps most famously illustrated in Jeremy Bentham's "felicific calculus"—appeal to our urge to simplify life. And, moreover, to make things *better*. Unfortunately, simplification is not necessarily the answer to our problems. "If we're talking about ethics," she comments, "I certainly do think that we need to resist the sort of simplification which naturally does occur to great reformers, in which they say the thing that really matters is X."[38] Quite the contrary. "What you don't have to say," she remarks, "is there has to be a single formula, a single way of arguing. Human life is complicated and it's no good to try to boil it down logically to something simpler. It isn't chemistry."[39] She continues, "There is no sort of simple account to be given of morality, which would say, 'Actually it's only like a machine.'"[40]

Midgley argues that issues serious enough to warrant the term "moral" have many facets and many partial solutions.[41] She embraces this complexity and insists that moral philosophy actually requires and thrives on the plurality, rather than reduction, of values.[42] Where others see singular answers, values, and solutions, Midgley, in other words, sees many. Animals, for instance, matter to us not just because they are subjects of a life (rights theory) or because they are sentient and have interests in not being harmed (utilitarianism). As valuable as these insights are, we might also argue that animals matter because we are deeply connected to them: affectively, socially, and evolutionarily. They are, in this way, fellow members in the community of life and have certain claims on us much like other community members (see "Animals").

In this way, more, rather than fewer, conceptual schemes are needed to capture the complexity of life and morality, even in relation to commonly shared moral issues. "If somebody asks," for instance, "'Why should we bother about our children?' or, 'Why have art?' or, 'Why, really, shouldn't I kill you?' many people will be somewhat stuck for an answer. This is not (as cocky graduate students tend to think) because these people don't know the answer, but because they know *too many* answers."[43] These "many answers" grow out of a range of factors—from particular contexts and experiences—not necessarily from broad, universal abstractions (though these can certainly help at times). Midgley is thus more amenable to pluralism because it goes the farthest in situating moral issues in the diversity of life, allowing for a range of potentially complementary values and claims. As she writes in her memoir: "Like William James, I go for pluralism."[44]

Midgley thus leans toward pluralism in seeking pragmatic common ground for reconciling and accommodating principles, insofar as possible. Various principles have their appropriate roles and capture aspects of morality, rather than the whole. For example, individual animals—domestic or wild—are indeed valuable, but what do we do when they undermine the integrity of the ecosystem that supports them and many other animals? Questions about how to handle these situations can result in clashes in principles—for example, between animal liberationist and ecological holist notions of value (see "Animals" and "Nature"). Oftentimes, though not always, our aims and goals in such polarizing scenarios might be brought together in more or less successful ways.[45] Incommensurability might arise, in the end, since some principles and conflicts can be reconciled better

than others. There are, in this way, real clashes that require difficult
thinking and force us into choosing the *lesser of evils*. (Do certain
animals need to be culled or should the hillside be allowed to erode?)
There is not, she comments, "any sort of quick recipe for dealing
with these conflicts but we are doing it all the time."[46] People can
and do work through moral issues with relative success, devising
and revising priority schemes and solutions. In turn, general sets
of standards and rules arise for working out problems over time,
though the difficult work of negotiating values and principles is never
complete.

Given this orientation to the diverse elements of our moral
systems, Midgley challenges systems and theories that simplify or
otherwise neglect important aspects of our moral landscape. She is
typically concerned with theories that amplify particular *categories* of
morality, most notably duty, rights, justice, equality, and freedom, to
the virtual exclusion of others. Each of these categories is important,
though some may include more confusions and conceptual tangles
than others. She is concerned not only with a good deal of the
conceptual tangles within these categories, but also with how
overemphasis on a chosen area of morality can overshadow things
outside of its boundaries. Individualist accounts of morality, in
particular, tend to overlook the central importance of relationships
in human life.

Sociality and others

For Midgley, morality "grows out of" our social nature in a Darwinian
context.[47] As such, the moral self cannot be conceived of in isolation,
but rather in the context of its many relations (see "Human Nature").
Yet, this intrinsically social context can be neglected in certain
conceptions of morality, especially in more individualist strains. There
are good reasons to elevate freedom and individuality: primarily, to
protect individuals from the very real historical abuses of collectives,
be they the state, social institutions, or society as such. However,
in doing so, Midgley warns that we can over exaggerate the ideals
of freedom or liberty and take these as *the* guiding principles in life,

much to the neglect of our bonds and interdependencies with others and the nonhuman world.

It can certainly seem that we are not all connected and interdependent. The steady bustle and occasional chaos of society, combined with our transience and the regular transformation of our social worlds, may leave us feeling as though we were flung together into a cement mixer rather than growing on the same tree.[48] It is hard not to feel isolated at times, even in a world increasingly connected through social media and other digital means. And, not only do we often feel disconnected but often we may wish to be disconnected. The competing interests and demands of a wide range of relations and institutions may make isolation and separation a desirable option, even if only temporarily. We do need space, solitude, and, ideally, personal time to process things and decompress from the constraints, demands, and stresses of family, friends, work, and so on. Yet, Midgley reminds us, whether we feel or wish to be disconnected, "*Society, however confused it may be, still does incorporate all of us*."[49] We are part of the whole and its wide network of relations. Apart from isolation in the wilderness, we cannot escape this reality.

Our social (and ecological) landscape is our setting and lifeline, however much competing visions of social isolationism may otherwise suggest. Midgley argues that atomistic visions—which abstract individuals as self-contained units—are, at a minimum, misguided and distracting. Though we are not identical with or subservient to society—as extreme communitarian or collectivist approaches may invoke—we are still part of society and cannot conceive of ourselves in utter isolation. We are parts of whole ranges of collectives: families, communities, ecosystems, and so on. Here we are shaped by these various relationships, groups, and entities. This does not mean we are completely passive members of these communities and collectives. As free beings, we also shape them in a dynamic interplay of mutual construction. We cannot exist apart from these worlds, but we are not wholly determined by or subservient to them. It is possible to create a balanced picture of our many relationships and dependencies without losing ourselves as distinct individuals.

Midgley argues that approaches emphasizing freedom—including such diverse figures as Hobbes, Nietzsche, and Sartre—typically do not mediate this balance. There are important values within this preoccupation with freedom—independence, self-reliance, autonomy, choice—that should indeed guide how we see ourselves and others. Yet, these are not the sole or primary values in human life. The danger here, Midgley warns, is that this individualist emphasis can subtly shift our perspectives from helping others to helping ourselves alone.[50] There are many situations in which people really *must* help themselves—abusive, asymmetrical, and self-denying relationships, for instance, not to mention the everyday ways in which personal boundaries are crossed and violated (e.g., workplace harassment). Midgley's target is rather the abstract and universal ideal of the free, rugged individual. It is not only a heavily masculine conception (see "Gender") but also an unrealistic and potentially damaging one. It can contribute to notions that we are not here to help each other and collaborate, but to tolerate or, in its more egoistic variations, compete with each other for limited resources (including freedom). Rather than pursuing the common good for ourselves and others, bound together as "members one of another," we bounce off of each other, like billiard balls, each pursuing our own self-interest while momentarily colliding here and there.

Midgley traces this billiard-ball mentality back to Hobbes's egoistic conception of human nature and social relations, which tends to frame our relations with others as a series of readily dissolvable contracts formed in pursuit of self-interested ends. This mentality, she argues, is still influential today in various forms, including various political and economic theories, underlying both the ideals of the free market and distortions of Darwinian theory. Yet, Hobbes is not alone in overlooking or downplaying the nature and importance of meaningful relationships and bonds. Philosophers from the Stoics to Kant—many of whom, Midgley notes, were celibate bachelors—regularly display this strange negligence.[51] Bonds are regularly portrayed as weaknesses rather than strengths, or as dangers that may bring more pain and sorrow (thus the Stoic advice to not become too attached to the well-being of others). Midgley faces the other direction and reminds us that our need for others "points outward to all the riches around us, the great stores of *otherness* in which we need to live." She continues: "Of course, our dependencies are dangerous, but who wants to live safely like a billiard ball or a doll that never leaves its package?"[52] The impression from these approaches, however, is that we must become mature and take full responsibility for our lives, relinquishing all the bonds that stand in our way. Midgley's position is that, though the acceptance of responsibility and self-direction is important, it is not incompatible with our factual dependence on others and our need for meaningful bonds. There are ways of balancing this picture, and philosophers have again followed their visions too far away from the reality of our social nature and needs.[53] "I am suggesting," she challenges, "that this extreme individualism is itself just a local and limited point of view, like other cultural world-pictures. . . . It is one of the many partial visions that we must use in our attempt to forge a workable worldview. There is nothing specially grown-up about treating it as a final revelation."[54]

Cultural considerations

Thus the social and relational elements of morality must not be overlooked in pursuit of enticing visions of individual freedom. But the dynamics of our collective enterprises, of the values and priorities that coalesce in our various contexts, are not without certain challenges. Most notably, we are faced with conflicts and challenges that can arise in an intercultural context—cultural relativism—as well as the triumphs and frustrations that can arise within cultural value systems themselves—the dynamics of social change. Individuals interact in various ways, but so do individuals and cultures, which results in the dynamic creation and revision of our moral visions.

Is it all relative?

The topic of cultural relativism opens into a whole range of wider questions in moral philosophy. For Midgley, much of the trouble regarding relativism stems from conceptual tangles related to skepticism and moral judgment. Skeptics, in general, rightly caution us from having too much confidence in our faculties (see "Science"). We are simply unable to have all the facts all the time in making judgments. Overconfidence in our abilities can result in rash judgments on a whole manner of issues as well as self-righteous (or worse) postures toward others. This moderate skepticism is thus useful in reminding us of our limitations and tempering our more ambitious moral systems. However, Midgley argues that skepticism about the inability to have *perfect* knowledge of all moral variables does not entail that there are no clear grounds for moral judgments. This all-or-nothing approach to the justification of moral judgments relies on unrealistic standards of knowledge (knowledge as complete and perfect *or* incomplete and useless). Too often, Midgley argues, skeptics take this overly ambitious route, obscuring or denouncing moral judgments in favor of relativism.

Moral judgments can and should be well-formed, clearly expressed, and publicly defensible. This difficult and involved process is often crudely and hastily done and looks more like moralism than moral philosophy. Unjustified moralism then solicits reactionary claims: How dare you make such judgments about *x* or *y*? Though this response to moralism is natural and well-intentioned, it encourages a climate of extremism between ill-formed moralistic judgments and counterbalancing claims that no such judgments can be made at all. Both extremes miss the important ground in between. Midgley argues that poor judgments are indeed the enemy here and that the emphasis should be placed on judging *better* rather than claiming that judgments must be abandoned altogether. This process of accountability entails argumentation and contextualization in the whole of life:

> We can . . . be expected to give those reasons from a system, however rough and incomplete, with which the rest of our lives coheres, and which is understandable both to outward and inward questions. If these reasons are grossly unsatisfactory, we then lie under an obligation to change the judgment. We cannot simply point to the rule-book, or lay our hands on our own hearts, toss off our wigs and go away.[55]

Judgments, much like skeptics argue, should not be haphazardly thrown about, but rather grappled with and justified in the context of our value systems. Here, again, we map our moral landscapes and articulate reasons for why certain worlds and certain ways of living are better or worse. Rather than outlawing judgment completely, which seems to be impossible (even for the skeptic), Midgley argues that we should rather "see to it that we judge fairly."[56]

These issues regarding moral judgments are relevant to cultural relativism, which Midgley terms "moral isolationism."[57] Moral isolationists, in brief, argue that it is always wrong to make moral judgments about other cultures. Midgley chips away at this position in several ways. First, it falls prey to the general problems of moral skepticism mentioned above: judgments *are* formed about cultures while officially prohibited by this position. Cultural relativists claim that we can never understand another culture enough to make moral judgments about it.[58] The problem, however, is that this position carries with it a whole connected set of judgments about other cultures, judgments which are important and to be taken seriously. For instance, one might claim that certain cultural values should not hastily be dismissed—there is probably a reasonable explanation in context for why cultures elevate etiquette or honor or the collective or the individual. Or, one might claim that cultures should be respected and tolerated because of their worth or dignity. Each culture, like one's own, should be respected for what it is—that is, if our culture is important and should be respected, so also should theirs (the cultural parallel to the enlightened ego). These, of course, are warranted judgments, but they are important *values* that require some type of justification. For Midgley, claims for respect and toleration do not preclude judgment—for instance, on female genital cutting or torture or patriarchal structures. Rather, these claims are parts of an argument that judgment should be done in *better*, more cautious ways. Attention should instead be given to critiquing "particular *ways* of judging—to bias, narrowness, chauvinism, lack of imagination, and willingness to limit individual freedom for the sake of Society" rather than abandoning intercultural judgment altogether.[59]

Midgley vividly illustrates this critique in her discussion of *tsujigiri*, the ancient Samurai practice of bisecting a "chance wayfarer" in testing the effectiveness of a new sword.[60] Against the charge of brutality, the cultural relativist may object that one outside ancient Samurai culture cannot make such moral judgments. At face value, this would simply be an assertion of the central tenets and values of relativism and the conversation would simply *stop* there. Yet, objections to judgments of brutality are typically the starting point. The argument usually continues, perhaps justifying the Samurai in more positive ways—for instance, praising the ancient ideals of discipline, devotion, sacrifice, and the common good in historical and cultural context. They might further argue that we can learn something from these practices, positioning this culture in judgment of our own. What is happening in these arguments is not an outright ban on intercultural judgment but contextualization, comparison, and critique. The relativist is actually illuminating how people are quite capable of understanding other cultures and making judgments in criticism or defense of certain values. Cultures are complicated, but they still involve values that are more or less intelligible to other people and cultures.

Cultural tangles

The need for cultural criticism or change, either in one's own or other cultures, is an important dynamic in this conversation. On the one hand, the meanings, symbols, and customs of cultures are central in shaping its inhabitants, and vice versa. This is a valuable insight of cultural relativism.[61] Yet, the sacrality of cultures is not absolute, and they may, at times, need to change. No culture is without its problems or above the fray. When this need arises, people must be able to look both within and outside of their culture for much-needed perspectives and insights. Here relativism, for all its insights, falls short because it blocks the possibility of this intercultural critique. "Morality demands," Midgley writes, "a wider horizon of possibilities than any existing society can supply. It is not just a matter of customs but of spiritual insights and ideals. Its field of reference is the world."[62] Sometimes people inside of a culture have the best perspective for critique—or praise, as both are useful—but sometimes those outside of a given culture can offer important ideas and values that would not have been seen otherwise.

Speaking of who is "inside" or "outside" of a culture, however, highlights a further problem in cultural relativism, which is the tendency to separate and reify cultures. This is something of the cultural parallel to social atomism, isolating cultures as distinct, complete units unto themselves. Midgley argues, accordingly, that cultural atomism or isolationism is as untenable as social atomism— neither we nor our cultures are so sharply divided and sealed. This is especially apparent in an increasingly global world of numerous overlapping and interconnected cultures and customs. Historically, this interconnection is certainly the case as many factors—colonial, economic, religious, and so on—regularly exerted their influence on the shape and makeup of any given culture, for better or for worse. Cultures regularly intertwine, and the boundaries of particular cultures are fuzzy and regularly shaken up. People often transcend cultural boundaries, straddling multiple cultural backgrounds and influences. Indeed, the "rules" of who is "in" or "out" of any culture—along with its many subcultures and sub-layers—are complex, multifaceted, and regularly contested. Cultures do, no doubt, differ in significant ways—recognizing this difference is essential in respecting cultural integrity—but, Midgley writes, "they differ in a way which is much more like that of climatic regions or ecosystems than it is like the frontiers drawn with a pen between nation states. They shade into one another. And in our own day there is such continuous and all-pervading cultural interchange that the idea of separateness holds no water at all."[63] These considerations are all the more challenging when we take notice of the various values and value systems within cultures themselves. Which systems are prioritized? Which are marginalized? How do power dynamics and other factors influence these systems and considerations? What about more recent revisions of traditional values within the same culture? To assume that one set of values typifies or characterizes entire cultures threatens to reduce the variability *within* cultures.

There are good reasons why moral isolationists seek to maintain and protect the distinctions of cultures. This, in large part, stems from an effort to shield cultures, especially nondominant and marginalized ones, not only from unreflective moralizing but also from the historical power differences between cultures. This is especially significant in consideration of the lingering impact and modern-day manifestations of colonialism and cultural imperialism. Many cultures have been oppressed and irreparably damaged by insidious outside political and economic forces, often under the banner of "moral superiority." This chauvinism is an ever-present threat to cultural integrity and should be resisted. Anthropologists, religion scholars, and other cultural theorists and critics have gone to great lengths to show the uniqueness and variability of cultures, along with what is valuable and worth preserving around the world. Harms to these traditions harm us all. Yet, Midgley warns that we must be wary, in the process, of making cultures out to be so distinct as to have no common ground or intelligibility. Here we run the risk of "othering" and fetishizing different cultures to the extent that they are virtually indecipherable and even "pseudo-speciated."[64] Cultures, in all their dramatic variability, are still outgrowths of a common human nature. We are all culture-building animals of the same species, which reinforces our unity through diversity. We should not assume that all people are the same, but we should also not assume that we are completely alien to each other. There is a sensible middle ground in which cultural diversity and integrity can be respected and appreciated while celebrating our common humanity and values.[65] In this, it is conceivable that we can all contribute to the give and take—the mutual learning and critique— that can constructively influence social change.

The possibility of change

Though cultures shape us, we also shape them. Cultural values and priorities, in this dynamic, reflect our needs and desires. As such, values are less fixed and unchanging than they may seem to be. Some values and priorities—liberty, individual expression—may be more central than others—equality, the common good—depending on the culture. There are usually reasons, some more explicit and defensible than others, as to why certain values are prioritized over others (usually phrased in terms of goods and harms). Social change is always possible because of the public nature of morality, even if societies are typically sluggish or downright intransigent in actualizing the values elaborated in moral discourse.

Most cultures have vast internal resources for change—a wellspring of ideas and criticisms forged very often through conflict and in times of pressing need—alongside the external resources and criticisms of other cultures.[66] Social change involves navigating this wider map. What can we change? How fast can it change?

Why are things like this? What can we do about it and what are our resources? Here we locate our current value systems and work toward the changes that are possible and desirable. Much change is urgent—radical injustice, gross inequality, asymmetrical power relations, and so on—while other changes may be less pressing. Some problems, such as climate change, may be prioritized because they exacerbate existing problems, while others, such as child abuse, may be prioritized because of the acute and irrevocable damages they inflict. Unfortunately, though we are quite capable of identifying problems and working toward potential resolutions, our collective judgments and behaviors often lag far behind and do not always track the urgency of these issues. This, Midgley notes, is because, though things can change, reform often requires more than just urgent need. For change to even be possible, the seeds of change must be already present:

> When the moral judgements of society do change—for instance, when people turn against cruel punishments, or begin to think that they ought to show some consideration for animals—this does not happen suddenly and alone. It is part of a wider change of attitudes, affecting parts of life not thought of as particularly moral matters at all—the kind of sympathies and revulsions that are felt, the customs that come naturally, the games that are played. But because this change is sweeping, for it to happen at all, there must be seeds growing in the culture for a long time, producing at first mainly uneasiness, conflict and distress. Even small moral changes are not easy and large ones are desperately hard; they go with total transformation of life.[67]

This seed metaphor highlights both the challenge and the promise of change. It involves the whole shape of our lives. Things can and do change.[68] However, change and reform—though occasionally rapid and abrupt—are typically lifelong and multigenerational projects. There are many seeds for better worlds, but they must regularly be (widely) re-sown and cared for to grow into the ideals that guide our actions. In this way, resources for change must be rediscovered and integrated with newer resources—heirlooms replanted, new varietals integrated—to cultivate better worlds.

Better and worse worlds

We will never likely create a perfect world. But we can cultivate a better one, and integrate the insights of current and past reformers. Cultures do have the resources to flourish, though these often require reexamination in the process—for instance, as when egalitarian reformers assumed that equality only really meant equality for *men* (see "Gender"). Unfortunately, reform is often slow and steady work that can frustrate us as we slam against the wall of reality. This is the paradoxical nature of the ideal. We set lofty standards on ourselves, yet we continually fail to meet these standards at the individual and corporate levels. We also seek practicality and waver between what is *possible* and what is *realistic*. This tension between idealism and practicality is a central challenge in morality, especially in applied ethics. We need standards that we can practically meet, but they must be worth striving for and, potentially, even unattainable. We need these ideals and visions because "realism" and "doing what's practical" often amount to implicit endorsements of status quo.[69] And we must continually reevaluate our standards—when we become complacent about the importance of certain values and ideals, the victories and accomplishments of those before us can be taken for granted and lose their hold. Here, again, the garden must be regularly re-sown and cultivated. In the following chapters, I explore Midgley's efforts to weed and sow the gardens of animals, nature, and gender.

4

Animals and Why
They Matter

*Central to thinking about animals is the idea of
community: a genuine unity of beings which are
important to each other.*[1]

Animals are a central element in Midgley's thought.[2] Her work, from
the start, focuses on the broad implications of the premise
that *humans are animals*. As such, one of her main priorities is
challenging the conceptual barriers that have been placed between
humans and the natural world. This is largely "destructive" work, as
she calls it, but it leads into a more constructive position on human-
animal relations. Humans are firmly embodied and embedded
creatures in Midgley's philosophy, continuous with nature and its
many inhabitants. Animals, in this context, are our fellow beings,
part of a "mixed" human-animal community. They are important
subjects, rather than objects or things, connected to us through
powerful emotional and relational bonds. This relational approach
was among the first in animal ethics to look outside of the prevailing
individualist categories of morality. By focusing on the emotive and
communal aspects of our relation to animals, she provided a shift
from conceiving of animals in primarily individualist terms (e.g., as
bearers of rights or interests). This relational angle connects with her
approach to the environment more broadly, which locates us within

the wider holistic context of the biosphere. Midgley's influential "mixed community" concept, in this way, envisions animals as part of the wider community of life on earth and significant among the many things that matter to us.

Barriers and dismissal

A number of barriers, conceptual and otherwise, have historically been placed between humans and other animals. These barriers, including our symbolic visions of animals, have significant impacts on how we think of ourselves and relate to animals (see "Human Nature"). Midgley argues that this animal symbolism, even in its more virtuous representations, tends to distort our conceptions of animal nature. It perpetuates images of animals as fictionalized characters— courageous lions and lazy pigs—rather than as real beings that share an evolutionary and, as she argues, social context with humans. Animals are typically symbolized as "other" beings in the world—as radically distinct and separate from humans. They are the inhuman or the antihuman, the stark contrast to our superior human nature and qualities. And, as negative representations of what humans are *not*, animals are easily seen as trivial objects, rather than subjects, in the world. In turn, they are either completely dismissed from or given a very low status in the moral domain. Because these representations and ways of thinking have real impacts on how we treat animals, Midgley argues that "we need new thinking, new concepts and new words, not . . . just about animals but about our whole relation to the non-human world."[3]

Both religious and philosophical sources have contributed to the barriers that have been erected between humans and the nonhuman world. The medieval Christian notion of the "Great Chain of Being," which imagined all beings from the heavens to the earth linked together in hierarchical relationships, has been especially potent. Humans, just below angels, have a special status above animals and the rest of creation. This imagery coincided with the biblical mandate (variously interpreted) to subdue and rule creation, and has long instilled a sense of divine right to rule over nature:

"Let us make humankind in our image, according to our likeness; and let them have dominion over the fish of the sea, and over the birds of the air, and over the cattle, and over all the wild animals of the earth, and over every creeping thing that creeps upon the earth." . . . God blessed them, and God said to them, "Be fruitful and multiply, and fill the earth and subdue it; and have dominion over the fish of the sea and over the birds of the air and over every living thing that moves upon the earth."[4]

This hierarchical vision of special creation and dominion has long persisted and not-so-subtly influenced the way that many have seen the nonhuman world. Humans are conceived of as specially created in God's "likeness," whereas other animals are not. Humans are, in turn, ontologically or essentially distinct from other animals and are their rulers by divine mandate. Midgley argues that this vision of the world—exaggerated by an emphasis on the singularly human nonmaterial soul—has historically legitimated all manner of anthropocentrism or "human chauvinism," characterized by arrogance, greed, and "a contempt for other life-forms."[5]

This religiously sanctioned outlook on the world funneled into the broader philosophical enterprises of the Renaissance and Enlightenment, where religious attitudes toward animals were largely integrated into humanistic and rationalist thought. The burgeoning fervor for humanist and rationalist pursuits, though not bad things in themselves, left little room for the recognition of animals. Animals were not on the humanist radar and, if anything, were typically brought into the conversation as a way to illustrate human dignity and value *over* other animals.

Midgley characterizes the dismissive attitude that characterized humanistic and rationalist philosophy as either absolute or relative.[6] "Absolute dismissal" is the attitude that rejects animal claims outright. This is exemplified by figures such as Descartes, who called animals "machines" or "automata," and Spinoza, who said "we may . . . use them as we please, treating them in the way which best suits us."[7] Here, animals are essentially and functionally equivalent to objects or machines for our use (reflecting, in Descartes's time, a developing preoccupation with machines and mechanistic ways of envisioning the world). As such, animals command no special moral importance, either theoretically or practically. Consideration or concern for animals is then rejected as supercilious fancy or, as Spinoza phrased it, "womanish pity" (a phrase which simultaneously dismisses both animals and women in one stroke).[8]

Arguments for absolute dismissal are rarely made—it is simply taken to be obvious that animals do not *matter*. As illustration, Midgley recounts the story of the trophy hunter R. Gordon Cumming.[9] In this example, Cumming shoots an elephant in the leg before having refreshments under a tree to admire his trophy. After some time (during which the elephant clearly suffers), he methodically shoots the elephant in various sensitive spots in an experiment to find the most vulnerable points. Eventually, the elephant is fatally wounded and put out of its misery. For the tradition of absolute dismissal, there is nothing directly wrong with the actions of Cumming in regard to the elephant. That is, there is nothing immoral in the treatment of the elephant itself—it might as well be a large stuffed animal. Rather, the wrongness, if there is any, is indirect. It is a matter of Cumming's defective character or perhaps in the potential that his callous actions toward the elephant may contribute to callous treatment of human beings.

In light of the brutality and callousness of certain positions of absolute dismissal, Midgley argues that the absolute position is less common than we might perceive, more often "professed than held."[10] More common is the position of "relative dismissal," in which animal claims are not dismissed outright, but rather given a "very low priority."[11] As such, animals are small blips on the moral radar and more of a distraction that "must come at the end of the queue, after all human needs have been met."[12] For relative "dismissers," the topic of animals is typically neglected, overlooked, or deprioritized in favor of more pressing humanitarian concerns. Thus, animals might matter in some way, but they demand far less attention than humans.

Midgley argues that the Western philosophical tradition contains traces of both absolute and relative dismissal. In actuality, more thinkers seem to dismiss animals relatively rather than absolutely. This is typically seen in the latent acceptance of animal subjectivity or value. For instance, she contends that even those who seemingly dismiss animal claims would likely draw a line on how long an elephant can be tortured and would distinguish this behavior from stuffed animal or machine torture. And, further, when one takes the position that elephant torture might corrupt character or contribute to the callous treatment of humans, there is an implicit assumption that the elephant is, to some extent, *relevantly similar* to humans and different from machines or other objects. Children are not allowed to torture animals for this reason: dissecting animals in the garden is not the same thing as disassembling machines.[13]

The scope of morality

Reformers from Voltaire to Jeremy Bentham resisted the seemingly arbitrary exclusion of animals from moral concern. Voltaire rejected Descartes's mechanistic position—"What a pitiful, what a sorry thing to have said that animals are machines bereft of understanding and feeling, which perform their operations always in the same way, which learn nothing, perfect nothing, etc.!"[14]—and Bentham famously challenged conceptions of moral worth that favored rational or linguistic capacities alone—"The question is not, Can they *reason*? nor, Can they *talk*? but, Can they *suffer*?"[15] This said, the main currents of moral discourse tended toward neglect and dismissal of animals. Assumptions of human exceptionalism manifested most forcefully in the possession of reason (and sometimes the soul). And, since animals, it was assumed, possessed neither reason nor soul, their moral worth was summarily dismissed.

The dominant accounts of morality also tended to focus on the favored categories of justice, duty, rights, and equality, among others. These categories were useful in furthering humanitarian reforms but typically functioned to narrow, rather than broaden, the scope of morality, often dismissing animals in the process. In trying to solve some issue of justice or rights or duty—often shading toward legal or political contexts—thinkers who may have otherwise considered animals worthy of some type of moral consideration were usually more focused on some other topic. This was usually a particularly narrow or technical area of moral discourse.[16] Thus Hume, in his discussion of justice, claimed that we should treat animals with "gentle usage," though they were not, as such, under the purview of justice.[17] Kant likewise claimed that animals were outside the bounds of direct duties, and that any duties we have toward animals are really duties toward humans (thus our duties to animals are indirect). Even though we may appreciate dogs and other animals, he writes that they are "there merely as a means to an end. That end is man."[18] Animals are means or things, not ends or subjects, and thus do not fit within the narrow circumscription of what makes a moral agent—a category only comprised of rational humans. They are, in turn, not directly within the scope of moral concern. This exclusion, though not always intentional, manifested in the notion of "*human* rights," a term which historically "has also the extra disadvantage of seeming to pre-empt" animal claims altogether.[19] Egalitarian literature, in a similar manner, positioned animals as the "unnoticed background" whose interests were placed below those pursuing human (male) equality. Being "treated like animals" is the ultimate insult to human dignity in this literature, though the brutal and callous treatment of animals is, at the same time, unquestioned and seemingly within the bounds of acceptable behavior.[20]

Midgley argues that the whole landscape of liberal moral discourse resulted in a confusing mess of conflicting perspectives on animals. This, again, is largely a result of reliance on the prevailing categories of morality (rights, duty, etc.) to the exclusion of other aspects or elements of morality. When the conclusions are that animals do not meet the *technical* requirements or scope of certain moral terms and concepts, the takeaway can easily be that *they do not matter*. Thus R. G. Frey's conclusion: "Animals cannot have rights because they cannot have interests. They cannot have interests because they cannot have desires or emotions. They cannot have desires or emotions because they cannot have the thoughts required for them. And they cannot have those thoughts because they cannot speak."[21] Midgley thinks it better to rephrase our discussions of morality altogether because the parameters set by this technical context are too limiting. Rights, equality, and justice alone, as useful as these categories may be, tend to obscure the scope of morality, leaving everything outside of their viewpoint in "outer darkness." She writes: "Whenever the spotlight picks out a particular moral area like this as central, things outside it tend to glide unnoticed into the shadows and be forgotten. Terminology, developed for central purposes, becomes unable to express them clearly."[22] Reliance on these terms tends to "imprison our attention in the area which has now become familiar."[23]

Lifeboat visions

In addition to the narrow limits of certain categories of morality, problems also arise from what Midgley calls the "lifeboat" vision of the world: a competitive posture of "me versus you" or "us versus them" in the competition for limited resources. Rather than view others as intrinsically related and "members one of another"—as integral parts of a greater community—it separates and isolates individuals as competitors for finite goods in a neo-Hobbesian state of nature (see "Human Nature" and "Morality"). Animals, in this vision, are competitors for limited resources (moral or otherwise) and thus threats to an already struggling lifeboat. Indeed, the lifeboat model has generally been used to limit our moral requirements to other *humans*.[24] We have our own problems and there is simply no room for animals.

There are, to be sure, serious considerations regarding the planet's finite resources and goods. There are, further, staggering injustices and inequalities in human affairs that must typically be prioritized. However, Midgley rejects this lifeboat approach to animals not only as disingenuous (again, the lifeboat model of ethics is typically used to limit our consideration of other humans) but as a distortion of our social and moral capacities. We are deeply social beings, and morality itself is grounded in our social nature (see "Morality"). Others are not mere competitors or "possible beneficiaries," as this contractual model assumes, but our fellow beings, members of our communities. Her relational vision, significantly, applies to animals as well. They are parts of our community—something we consider *serious* or *important*—and thus demand the care, concern, and attention that fellow community members typically require. Midgley argues that our world is much richer and more complex than the lifeboat vision allows. "We are not usually in lifeboats," she wryly observes, and such black-or-white ways of seeing the world intentionally limit our ingenuity and creativity while simplifying our social sympathies.[25] There is little room for compassion, generosity, cooperation, and our significant emotive capacities. Questions are reduced to prudential considerations of who is "in" or "out" of the lifeboat: You? Them? This group? That nation? Animals? The environment? It conceives of morality as a series of concentric circles in which the inner circles, emanating outward from the egoistic self, are prioritized in our moral considerations as a matter of simple calculation—in *or* out. Lifeboat approaches, in other words, abstract us from the wider context of life as a whole, forcing us to consider moral issues in terms of narrow and unrealistic visions of an egoistic war of all against all. They misrepresent and distort human life and morality and often preempt much concern beyond our species (other than prudential concerns for resources). Though competition and finite resources are real *aspects* of the world, to frame our moral landscape exclusively in these economic terms overlooks too much that matters.

Animal liberation

Moral visions of the world that threaten to dismiss animals either by narrow technical restrictions or lifeboat ethics have been resisted by animal ethicists since the 1970s. Animal liberationists have made significant gains in animal discourse by utilizing the same concepts and categories—typically rationalist individualist ones— that had ironically been used to dismiss animals. Peter Singer, for instance, forcefully demonstrated how the prevailing principle of equality, a concept so central in humanitarian reforms, is a matter of equal consideration of interests. If animals have the capacity to experience pleasure and pain—or, significantly, to suffer (drawing on Jeremy Bentham)—their interests should be considered alongside humans, since these are the measures by which we consider human interests.[26] Otherwise, we are not consistently following our principles—a rationalist taboo—and are erecting arbitrary moral boundaries. Likewise, Tom Regan and other rights theorists have put forward theories of respect that value animals intrinsically. They demonstrate how Kantian and other duty- and respect-based approaches to ethics arbitrarily draw the line of moral concern at certain human capacities and characteristics. However, when these capacities are more fully examined, they appear chosen as moral criteria simply because they are human qualities. Regan argued that many animals fulfill the category requirements of being a "subject-of-a-life," much the same as humans, meaning that they have beliefs and desires, along with "perception, memory, and a sense of the future, including their own future; an emotional life together with feelings of pleasure and pain; preference- and welfare-interests; the ability to initiate action in pursuit of their desires and goals; a psychophysical identity over time; and an individual welfare in the sense that their experiential life fares well or ill for them."[27] This robust set of subjective qualities demands that moral agents recognize and respect the rights of animals. Both Singer and Regan argue that, our moral principles, when openly and honestly examined, can and often do encompass more than our species and, when employed consistently, can be extended to include, rather than arbitrarily exclude, animals. To do otherwise would be "speciesist."

These and other influential welfare approaches have brought attention to animals and have, perhaps most importantly, prompted real-world changes in farming and research practices while igniting the animal rights movement.[28] Midgley praises Peter Singer, in particular, for drawing attention to animal suffering and challenging the inconsistency and arbitrariness of rationalist categories of morality.[29] This said, she notes that, in spite of these reforms, the extension of individualist concepts of moral worth to animals—interests or rights, for instance—also extends the limitations of these approaches. In focusing on individual animals as the unit of moral concern—as a being that suffers or possesses other psychological or emotional capacities—the wider communities and the various collectives in which animals live and rely on (ecosystems, etc.) are not fully accounted for. And, by employing rationalist categories and assumptions, relations, emotions, and other important elements in our moral landscape are overlooked. Midgley consequently argues that these liberationist approaches are valuable as particular *aspects* of animal ethics (see "Morality"), but not enough to cover the whole. They each play a certain role and do certain jobs better than others, but their approaches and solutions are not final. They focus on one particular category of morality—complete with its own conceptual tangles—and run the risk of spotlighting it to the exclusion of others:

> [Moral philosophers] have . . . repeatedly piled far more problems onto the concepts of "rights" and "justice" as well as "value" than those words can stand, instead of looking round to see what other notions might be helpful. (What about "importance"?) And they also have a tendency, dating from their most rule-bound epoch, to treat things as alternatives which are demonstrably not alternative but complementary—to dispute about whether we should use knives or forks instead of asking how we should combine them.[30]

Midgley, in turn, shifts the moral spotlight and draws attention to the importance of our continuity and relationality with animals.

Envisioning humans and other animals

Midgley's philosophy is one of embodiment and embeddedness—we are firmly part of nature and continuous with other animals through our evolutionary heritage (see "Human Nature"). Rather than hovering above nature (as links in an ethereal chain, for instance), we can see ourselves variously related to other animals on a horizontal, "radiating bush" of life.[31] In this context, "natural selection is no more, if no less, on our side than on that of the wasp or the tapeworm."[32] For Midgley, this lack of hierarchy and *telos* or direction in nature challenges traditional notions of the Great Chain of Being and human exceptionalism. While there is nothing *wrong* with taking human life and destinies seriously—something we could always use more of—she argues that this is no reason to hold "a contemptuous attitude to other destinies."[33]

In envisioning a shared world of (inter)related beings, Midgley pushes back against the Cartesian view that animals are simply machines. This prejudiced metaphysical view slams against the wall of decades of animal studies and thousands of years of practical experience and interaction with animals. To hold this view today requires virtual isolation from real animals and science as well as "some awkward intellectual gymnastics."[34] This is because we have experienced and discovered that animals, in varying degrees, possess similar nervous systems, social patterns, and other characteristics that are structurally similar to humans.[35] This does not mean that all animals, in their various natures, are *just like* humans. At least not in every relevant way. And, importantly, *likeness* to humans does not mean *better*, which is a tempting way to smuggle hierarchical assumptions into thinking about animals. Rather, the takeaway is that these relevant similarities and shared features point toward animals as subjects pursuing their own characteristic lifestyles, in which some things are good for them and others not.[36] As Tom Regan and Paul Taylor argue, they have a certain direction in life which can go better or worse for them.[37] This is not to posit a mysterious force compelling animals toward some cosmic destiny, but rather to zoom in to the lives that animals lead as real living creatures, as squids, hawks, groupers, and lemurs. While some animals, such as honeybees, may rely more heavily on fixed patterns of behavior, others, such as dolphins and wolves, demonstrate an expansive range of flexibility and adaptability, suggesting that the caricature of machine-like automata operating automatically in the world is more of a Cartesian fantasy than a scientific doctrine. Of course, the subjectivity and complexity of animals varies dramatically—slugs and whales, for instance, differ immensely—but the "spark," so to speak, of rich, subjective life is present in the world and shared by animals in different degrees. As such, Midgley argues we should recognize that humans do not have a monopoly on subjective and meaningful existence in the world. This, at a minimum, is a good reason (among others) to take animal life seriously.

In fact, we cannot help but take animal life seriously when we take *human* life seriously. As Midgley's view of human nature reminds us, we are indeed animals and cannot escape our nature (see "Human Nature"). This common ground with other animals—our shared natures and evolutionary heritage—makes comparative study and understanding possible. In this way, it is not unreasonable to extend the vast store of experience and wisdom we have of human life into more realistic representations of animals (and vice versa). The descriptive language is ready at hand in how we think of ourselves. Humans have inner and outer aspects—subjective perceptions and outer behaviors—that we constantly align with the world.[38]

We regularly employ descriptive language in describing someone's inner state by interpreting their outer states. "She is angry." "He is frightened." "They are in love." We may not always be perfectly accurate in our interpretations and attributions. "He was just tired." "She was actually more nervous than anything else." And sometimes we completely fail in understanding others. This said, our lifelong calibration of experiences with those of others is probably enough to give us a qualified confidence in our interpretations and interactions. And, in the case of animals—as fellow beings—we are quite often able to employ similar ways of understanding their motivation and behavior. "The dog is anxious." "The chimpanzee is frustrated."[39] "The octopus is trying to deceive." The same tools that have evolved for understanding other humans are likewise useful, to an extent, for understanding animals. Of course, this can be overstated and overstretched, especially in cases where the relevant similarities are too few. However, the relevant similarities of humans and dolphins and bonobos and wolves, for instance, are likely significant enough to interpret particular inner and outer states.

Red herrings

This ability to understand animals, however qualified and limited, challenges the skeptical insistence that the "subjective" feelings of animals are "quite hidden from us, cannot concern us and may well not even exist."[40] This extreme skepticism, characteristic of behaviorism, claims that we can only observe and understand outer states or behaviors and finds subjective interpretation dubious and unscientific. But while skepticism in interpreting states is always warranted, Midgley counters, our imperfection in this practice is not a strong enough reason to deny our ability to say *anything* positive about animals. Here, the charge of "anthropomorphism" is quickly made if any attribution of human emotions or subjectivity is made of animals. "I'd say the dog seems worried, but I don't want to anthropomorphize." "We might say this elephant is excited to see her old friend, but that would be anthropomorphizing." This "anthropomorphic clause," used as a shield against being "unscientific," comes with good intentions—we should certainly aim to be careful in attributing various states to animals and not make them out to be something that they are not. However, if things are framed in our shared evolutionary context, Midgley argues, anthropomorphism is only wrong when it *improperly* describes the emotion or feeling. Otherwise, it is, to an extent, sensible and appropriate to refer to corresponding states and emotions between humans and animals in this language.[41] The dog probably is worried and the elephant probably is excited.

What we should avoid is *humanizing* animals—portraying them as equivalent to humans in character, and so on—or pretending that there are no significant boundaries between us and other animals. Lions are not "kings of the jungle" and companion animals are not surrogate humans. This said, Midgley pushes back against the timidity or embarrassment of saying anything positive in everyday or even scientific language about the subjective aspects of animals. This, again, is more of a philosophical commitment than a scientific discovery and she wishes to remove the stigma of anthropomorphic language altogether as a red herring:

> This attack assumes that human language is invented in the first place not only *by* humans, but exclusively *about* humans—to describe them and them alone. Any use of it to describe any other being would then be an "extension"—a leap out into the unknown. But if language has, from the start, arisen in a mixed human–animal community and has been adapted to describe all beings whose moods etc. might be of general importance and interest, then that is the proper use of the concepts from the start, and no leap is needed.[42]

Anthropomorphizing language, then, is a defensible way to refer to animal behavior on many occasions and reasonable in light of our shared evolutionary background with other animals. Skeptics are right to caution against overconfidence in our capacities to interpret the subjective states of animals. However, to reject from the start that any interpretation is anthropomorphic and therefore invalid undermines the very faculties we use for interpersonal understanding of *humans* with success. These faculties might not be perfect, but they can be practiced, honed, and improved enough to reliably work. This, again, is a flaw of extreme skepticism: the assumption that knowledge must be *perfect*, and, if it is not or cannot be, it is useless and not worth pursuing. Here, a more "moderate skepticism" can realistically balance our interpretations of others against our tendency to occasionally be mistaken. More drastic maneuvers are unnecessary and less scientific than radical skeptics suppose.

Fuzzy boundaries

Midgley argues, moreover, that if we take our evolutionary context seriously, many of the barriers that have been placed between humans and other animals become fuzzier and much less secure. What distinguishes us *among* the animals is less distinct and final than is often assumed. There is certainly real human uniqueness—for example, we are quite distinctive in our measure of self-awareness, the will or effort we can exert on the world, and the extent to which we can reflect on our conflicting motives and desires, among other things (see "Morality").[43] This said, these and other qualities are typically distributed on a spectrum binding humans and other animals. Clusters of structural properties—physiological, behavioral, psychological, and otherwise— make up any given species, including humans. Given the diversity of species on the planet, there are, in turn, virtually limitless numbers of these species-making clusters, some more overlapping than others. In the case of humans, we have a remarkable collection of properties, some of which locate us in close proximity with certain animals and more distantly from others. Our social structure and complexity, for instance, is similar to whales and dolphins, and it is well known that our genetic makeup overlaps with a range of animals, most notably chimpanzees and bonobos.[44] Of course, we still differ in many relevant ways—chimpanzees are less social than humans, for instance, and we share much less with fish and reptiles than with birds and mammals. In this dynamic understanding of animality in context, the characteristics that have traditionally been trumpeted as our distinctive excellences— thought, language, culture, consciousness, and the like—have real significance but are firmly located within the collection of properties that makes any species.[45] We all come from the same source.

Midgley's analysis of species and the species-barrier is subtle, complex, and wide-ranging. In examining the continuity of humans and other animals, she resists the temptation to dismiss the species-barrier as trifling or insignificant. Species lines certainly are less distinct than often assumed in the Western philosophical tradition and the popular imagination. Yet, these species boundaries cannot, as some animal advocates suggest, be dismissed as superficial or illusory. Here, she has in mind the generalized claim that preference for one's own species is always arbitrary and immoral.[46] She agrees with this, to a certain extent. However, her overarching argument is that distinctions between animals are no trifling matter. The continuity of life on this planet unites all animals, in varying degrees, but it does not make them identical. Awareness of species is ultimately essential to properly understanding and respecting them. To do otherwise is to engage in a form of "patronizing thinking," which flattens out the integrity and distinctions of different animals and overlooks that we are all "beings of particular kinds."[47]

The species-barrier then is real, important, and should not be marginalized. *But nor should it be exaggerated.* However real and important it is, it is also flexible and semipermeable. This is the case for whole ranges of animals, whose continuity and overlapping boundaries allow for numerous ways of interacting and living together. Relationships and interactions between species vary by context. Life can certainly be cutthroat, destructive, and volatile. Yet, this is not the only law of nature. Interspecific toleration and even symbiosis—oxpeckers eating ticks off the backs of rhinoceroses and cleaner fish removing parasites from larger fish—are well-known examples of the highly interactive symbioses that can happen in the wild. And, in domestic settings, we regularly see heart-warming stories of animal adoption and bonds—dogs taking in ducklings and horses that become attached to barnyard cats. These special relationships, some more recognizable than others, are characteristic of the wider dynamics of a mixed-species world. Animals are generally aware of and in communication with others as a matter of course. This interspecific interaction, of course, should not be mischaracterized as harmonious and utopian. But neither should it always be exaggerated as a war of all against all: "nature, red in tooth and claw." Our images and visions of the nonhuman world must be *balanced.* To paraphrase Raymond Williams, ideas of nature must include both the cyclone and the butterfly.[48] Relations can be quite destructive—carnivores, omnivores, and parasites require nutrition—but this is not the total picture. There is also significant tolerance, cooperation, and even special bonding that connects this wider community of life.

The mixed community

A more balanced and comprehensive vision of other animals may lend us surprising insights about what we consider important. Perhaps the most significant and fascinating aspect of this wider map of species and interactions—this mixed-species community—is how humans operate within it. We are bond-forming social animals, but these bonds are not exclusive to our own species. While we form intense short- and long-term relationships with other people, our bonds and relations transcend species-barriers and give rise to a mixed human-animal community: the "mixed community." Here, relationships of various kinds, degrees, and distances characterize our deep connection and involvement with other animals. Companion animals are classic examples of the relationships and small communities that coalesce between species: we essentially adopt other animals and see them as parts of our families (cats, dogs, horses, etc.). This phenomenon is a fairly natural outgrowth of our tremendous social and emotive capacities and impulses. Our shared traits with these and other animals, along with our empathetic and sympathetic powers, have placed us in a unique position to relate to and understand animals in impressive ways. We may not necessarily form bonds with wild animals in the same way we do with domestic animals—though this can and does happen at times with elephants, raccoons, dolphins, and so on—but our common heritage and connection in the community of life is real and important.

Midgley argues that our community with other animals—and the sympathy, compassion, and care therein—must feature more prominently in our moral visions. However, the relations and emotions that characterize this community are easily overlooked, downgraded, or rejected in approaches that favor sweeping moral principles. This is what, Midgley argues, has typically happened within the rationalist individualist tradition. Moral principles are meant to be binding on everyone (or "universalizable"), regardless of how one *feels about* or *relates to* others: one should always equally consider the interests of sentient beings (beings that can suffer) or respect the rights of a subject-of-a-life. This approach makes sense in devising compelling and consistent reasons to treat animals better in accordance with our general notions of moral worth. Here, animals are forcefully shown to meet at least some of our prevailing standards for moral consideration (possessing certain capacities, for instance). Yet, Midgley argues, this account is incomplete:

> What makes creatures our fellow-beings, entitled to basic consideration, is not intellectual capacity, but emotional fellowship. And if we ask what powers can give a higher claim, bringing some creatures nearer to the degree of consideration which is due to humans, what is most relevant seems to be sensibility, social and emotional complexity of the kind which is expressed by the forming of deep, subtle and lasting relationships.[49]

If we give due weight to all the types of bonds and connections that matter, we will be in better position to take animals seriously.

Animals are significant members of our communities, locally (in our homes and neighborhoods) and globally (in our ecosystems and planet). All of these connections and communities matter, from the particular and local to the general and global. However, localized human-animal communities—typically the "domestic" settings with which we are most familiar—are especially unique. These communities are manifestations of the wider human impulse to connect to the world around them. "All human communities have involved animals," she observes, and it is "one of the special powers and graces of our species not to ignore others, but to draw in, domesticate and live with a great variety of other creatures."[50] Such domestication was achieved largely because many animals (such as dogs, horses, pigs, etc.) share a number of social and emotive characteristics with us. They were, in turn, able to form bonds, understand social signals, learn to obey particular persons, and so on. These shared traits and behaviors made possible the historical development of complex human-animal communities that have taken innumerable shapes and forms over time and space.

Though animals may not be equivalent to "persons" in these communities—a term Midgley argues is too loaded with legal baggage—as members of our mixed community, they are certainly fellow *subjects*, not objects or things. The simplistic Kantian antithesis of "persons *versus* things" is unhelpful and obscures our recognition of animal subjectivity.[51] We have long recognized the subjectivity of animals—indeed, animal cruelty paradoxically underscores this fact.[52] Cruelty involves a belief that something is a subject capable of suffering. Dogs are kicked and horses are beaten not because they are things (like machines and stuffed animals), but because they are beings that feel and experience pain in a significant way. In other words, belief in animal sentience is essential "for exploiting them successfully." In fact, Midgley points out, "Exploitation *requires* sympathy."[53] Abuse and cruelty to animals is an outcome of our ability to understand and relate to the "inner" as well as the "outer" states of other animals, coupled with the tendency to devalue or disregard these states.

Yet, we need not dwell on the negative. Genuine interspecific love and care are pervasive, from intense bonds of companionship—many of which are held to be as important as human bonds—to altruistic acts toward animals (rescuing injured animals, animal welfare charities). Instances of cruelty do arise, but these are unfortunate episodes in a long history of coexistence. In fact, animals are typically seen as significant features of human life. They are featured prominently in our imagination, identity, and social worlds. It is hard to even imagine life without animals, be they companion animals, songbirds and squirrels, or characters in books and films. In this mixed-species context, in which our very language is reflective of our history and coexistence with other animals, most of us are imprinted by interspecific sociality from a young age. We crave animal contact from our youth, and it is a foundational element of our early lives along with song, dance, and play.[54] Bonds with animals work alongside our bonds with people as part of a "full human life."[55] Human and animal worlds intermix and overlap in powerful ways, especially in our youth, which influences how we view life from the beginning:

> The species-barrier, imposing though it may look, is rather like one of those tall wire fences whose impressiveness is confined to their upper reaches. To an adult in formal dress, engaged in his official statesmanly interactions, the fence is an insuperable barrier. Down below, where it is full of holes, it presents no obstacle at all. The

young of *Homo sapiens*, like those of the other species present, scurry through it all the time. Since all human beings start life as children, this has the quite important consequence that hardly any of us, at heart, sees the social world as an exclusively human one.[56]

This childlike wonder and curiosity motivates us, throughout our lives, toward this fascinating "otherness" in the world, human and nonhuman alike. That is, we are drawn to and capable of appreciating the world and its many inhabitants in affective and moving ways. In this intra- and interspecific community, love for animals and the nonhuman world complements love for humans. And, by caring more, we widen our horizons—we need not limit ourselves to small spheres or circles of concern. Midgley's point here is that, though our emotive capacities do have genuine limitations—we cannot love and care for everything all of the time—the boundaries of these capacities do not and need not stop at the species-barrier. "Love," she writes, "like compassion, is not a rare fluid to be economized, but a capacity which grows by use."[57]

Hyper-rationalist approaches to ethics intentionally abstract out or bracket off the emotive capacities that underlie our connections to others, including animals. For Midgley, this move is incomplete, since both thought and feeling come together in moral deliberation as "aspects of the same process."[58] Emotions and their interconnected thoughts—for instance, indignation as feelings and thoughts regarding proper outrage—have roles to play in a more integrated and comprehensive approach to animal welfare. Emotions, in other words, are entangled with concepts and thoughts. These "scruples felt" power and animate morality and are, so to speak, the "raw material" for moral systems.[59] They are what drive our moral visions. These visions are central to the principles that we derive.

Animal claims

Thus Midgley draws attention to the importance of our relationality, intersubjectivity, and emotional connection with animals. Yet, while focusing on these more relational elements, Midgley is quick to stress that nearness in relationship status (family, friends, etc.) or physical location (neighbors, nation, for example) is not the only important factor with which we should be concerned. Nearness *is* a key factor in our moral relations and must be taken seriously—close personal bonds and relationships are a central part of life—but there are other claims that can outweigh those nearest to us, such as those of humans and animals more distant from us (in addition to claims from the wider nonhuman world). "The moral universe," she writes, "is not just a system of concentric circles, in which inner claims must always prevail over outer ones."[60] For Midgley, no system of concentric circles helps us fully understand and decide on difficult moral dilemmas. This approach is too centered on the *central* circle: the ego. She argues that this model is too simple and restrictive, since our concern and capacities easily spread beyond the more immediate circles of self, family, neighbors, and so on, reaching out to various others (human and nonhuman alike) with ease. Theories that aim to identify a "simple formula" with which one can assess competing claims "make the job look simple" and "can only deceive us."[61] Instead, we must work out maps of overlapping *types* of claims. This is not to "fix priorities," but to mark our recognition that "relatively isolated claims" sometimes prevail over those nearer to us.[62]

The types of claims that we might map are various. They include, but are not limited to, claims of fellowship, kinship, justice, special need, special responsibility, prudence, gratitude, admiration, and wonder, which all arise in varying degrees and circumstances. These shifting and overlapping claims do not perpetuate sharp divisions between "us" and "them" as concentric circle models do. That is, these claims can coalesce for any number of beings and even entities, both near and far. They also highlight that there is no straightforward species-barrier "exchange rate" that we can use to calculate and adjudicate claims directly.[63] Circumstances "on the ground," rather than sweeping principles, influence what types of decisions we should make. We often have special responsibilities toward domestic animals, for instance, because we have removed them from their natural habitat and taken them into close fellowship in our social worlds. Likewise, there are sets of overlapping claims (special responsibility, prudence, wonder, etc.) that might arise when our actions threaten a whole ecosystem and its many inhabitants. Clusters of overlapping claims in context largely determine what considerations are necessary in weighing short- and long-term actions and policies.

What develops for Midgley is a comprehensive approach to a wide array of complementary and interrelated ethical issues, in which animals and the natural world are not isolated but integrated into the broader framework of our moral deliberation. Concern for animals blends quite seamlessly into concern for humans and matters of social justice, family, class, race, gender, and so on. These issues are all tied together as things that matter to us and are central to our lives. An effective approach to animal welfare, she proposes, must connect animal issues to these wider sets of issues and claims and involves bringing together flexible "sets of principles for relating them."[64]

Domestic and wild

Midgley's mixed community concept is part of a generalized picture our significant evolutionary, social, and ecological connections with all life, which manifests in specific and dynamic ways over time. These manifestations may be more clearly recognizable and distinguishable in certain places and times—especially the local and domestic—though the wider landscape of our community, envisioned in this way, stretches broadly, much like we imagine the wider human community. As such, the mixed human-animal community that we live in does not necessarily terminate at the troublesome and oft-contested line between domestic and wild animals. Animals can be said to fall along something of a domestic-wild spectrum based on a number of social, ecological, and conceptual considerations. Dogs, feral cats, squirrels, elephants, songbirds, chimpanzees, sheep, gerbils, and dolphins are vastly different animals whose domestic-wild statuses are not always especially clear. This status, which is never likely clear-cut and final, is constituted by a range of considerations, including, among other factors, the diverse and complex roles and relationships animals have as family members and companions, nuisances, coworkers, predators and prey, symbols, characters, food, trophies, religious objects, research subjects, and so on. That is, there are many factors contributing to whether an animal is more domestic or wild or semi-domestic— "sort of half way in and half way out," as she puts it[65]—and differing contexts and circumstances influence where these animals land.[66] These distinctions are typically less *distinct* than we often assume.

Midgley does focus mostly on what we would typically consider domestic animals in discussing the mixed community, as this is the realm where our shared sociality with animals is most familiar. Yet, this connection with animals does not terminate at the domestic-wild borderline, wherever that may fall.[67] Domestic communities are vivid examples of the overarching human-animal community that we are part of, or the "genuine unity of the whole of beings which are important to each other."[68] Resisting abstraction, this greater community manifests in particular and concrete ways, both spatially and temporally. "A wide concept like the mixed community," she remarks, "is naturally going to dissolve into a lot of little ones."[69] Though pragmatic concerns and clashes no doubt arise, these diverse communities involve more than just domestic animals. Consequently, as a conceptual tool for envisioning our relationship with animals, the mixed community resists rigid separations between domestic and wild animals, though the factors contributing to these categories are certainly real and relevant.

Even though the terms "domestic" and "wild" might be relevant in particular circumstances—clashes between sheep and wolves, for instance—they do not always provide us with sharp, tidy boundaries for weighing and prioritizing the various types of claims that might arise with different animals in different contexts and communities.[70] We do generally treat wild animals differently from domestic animals, and this is well-warranted given the varying natures and contexts of different types of animals. We should probably not, for instance, bring a poisonous snake indoors or leave a chicken alone in the woods. Animals, again, have differing needs and behaviors, all of which must be considered in recognizing their dignity and, if possible, in allowing or encouraging their welfare. In other words, there *are* relevant differences and distinctions between certain domestic and wild animals that impact their claims and treatment, but we must clearly articulate what those are and, as best we can, frame their claims and needs accordingly without lumping all animals into simple categories.

This nuance is important because the sharp categories of domestic-wild have, at times, been utilized to take one set of claims more seriously than others. Midgley complicates this by placing domestic animals, wild animals, and everything in between on the wider map of our multiform communities. Early approaches to animals and the environment tended to gravitate toward the extreme poles of individualism and holism. Individualists mostly focused on individual animals, typically in a domestic context, as the primary unit of moral concern. Their approach was to extend existing individualist moral principles to animals (Singer and Regan) in a much-needed effort to curb cruel treatment and exploitation of animals for food, experimentation, and so on. Holists, stemming from the tradition of Aldo Leopold's influential land ethic, focused on collectives: ecosystems and the biosphere, as well as the many dynamic interlocking features and entities therein. This was in an equally important effort to curb rampant environmental destruction, degradation, and disintegration. Yet, for holists, animals mattered less as individual units of moral concern and more as species or as parts of ecological wholes. That is, this approach mostly valued individual animals in relation to their respective collectives while placing more value on the collectives or wholes themselves. Individualists became associated with the animal liberation movement and ecological holists with the environmental movement. Though these causes need not necessarily conflict, as Midgley argues, they often clashed over the relative value of individuals versus collectives, which often manifested in their respective emphases (or lack of emphases) on domestic or wild animals.[71]

Animals and the wider world

Midgley brings together concern for both individuals and collectives in her exploration of social and ecological claims. *Social claims* are the claims of sentient beings. She characterizes a sentient being as "one which can *mind* what happens to it, which *prefers* some things to others, which can be pleased or pained, can suffer or enjoy."[72] With social claims, we respond to the "spark of conscious life" in the world and prioritize its importance.[73] These beings matter as subjects for whom things can go better or worse *and* as members of our communities.[74] This said, the claims of our world do not stop at sentience. There are other claims that exist in things that are non-sentient—plants and trees, for instance, along with non-sentient animals, various types of collectives, and the wider features of ecosystems. *Ecological claims* are the claims we respond to on behalf of non-sentient beings and entities: forests, species, collectives, and the biosphere as a whole. Midgley argues that concern, care, and responses to claims can and do extend widely in this way, as we have a deep "capacity for interest in other, inanimate surrounding objects—plants and stones, stars, rocks and water—which extends our horizon beyond the social into the ecological, and makes us true citizens of the world."[75]

Midgley resists the idea that social and ecological claims *necessarily* clash—that one must value the interests of individual animals over collectives or vice versa. Some have argued along these lines, claiming that individual (social) and holistic (environmental) claims are inherently at odds with each other.[76] There is certainly merit in the different types of claims that are emphasized in the individualist and holistic approaches, because "concern for the whole environment gives no direct motive to oppose bullfighting, nor does humane concern for bulls directly forbid the proliferation of cars. These are distinct campaigns. Even if they seem closely connected and are often pursued by the same people, they differ widely in emphasis."[77] These distinct campaigns are particularly important in recognizing and drawing attention to historically overlooked and neglected sets of claims: battery chickens and polluted waters really do need attention and devoted advocates. However, it is also important to recognize that these causes "overlap widely" and do not necessarily entail "all-out conflict."[78] She puts this in typical Midgleyan fashion by analogizing: "If you are in favor of bread, as it were, it doesn't mean you *can't stand* butter."[79] Fanaticism, as she terms it, for either set of claims to the exclusion of other is too restrictive and oversimplifies "the moral scene." Social and ecological claims are typically continuous and the difference between them is not always so stark. Animals and their respective environments, domestic or wild, are too dynamically integrated and interrelated to have their claims rigidly separated. Most often, these interests and claims are intertwined and mutually reinforcing.

This said, though social and ecological claims are often harmonious and continuous, there are times when they do indeed conflict. Midgley observes, however, that this clash is "no more surprising than other clashes between different sorts of moral claims, and we have to deal with it in the same way, namely, by working out the best system of priorities that we can manage, and not going out of our way to intensify them."[80] In these situations, there can be irreconcilable conflicts that lead us not to perfect solutions, but to a choice between "evils." She urges us to consider the sliding scale of "better and worse" choices over perfect "one-size-fits-all" solutions. Here, we must consider the wide sets of claims and priorities involved while avoiding oversimplification:

> We are familiar with such clashes between other important ideals—between justice and mercy for example, or between all our duties to others and the duties of our own development. There is no clear, reductive way of settling who wins this kind of contest. We know that in these cases we can face a real choice of evils, and we then have to find some way of deciding which of these evils is, in this particular case, the worse.[81]

This is particularly challenging when humans are usually the root cause behind certain clashes—for instance, when species introduced by humans threaten the integrity of an ecosystem. Invasive species such as rabbits and feral goats can dramatically impact the local environment by overgrazing, crowding out native animals, and generally disturbing the food chains and wider dynamic mechanisms of an ecosystem. (Other prominent examples of disruptive introduced species include cane toads, Asian carp, Burmese Pythons, and European Starlings in the United States.) Yet, the many overlapping and clashing claims involved, and our attendant obligations, are not always so clear. Rabbits and goats, like other feral or introduced animals, have numerous claims. As sentient beings, their social claims include not only aspects of fellowship and justice but also special responsibility—doubly so, since humans both domesticated these animals and, further, re-introduced these animals to an environment! The local ecosystem, on the other hand, likewise has claims on us. This becomes quite complex because, as ecological holists remind us, the ecosystem supports many other organisms, including other animals that may have social claims on us. In turn, numerous other social claims may arise and overlap with the urgent ecological claims of justice, special need, and prudence, to name a few.

Blanket prescriptions regarding individuals and collectives will most likely not give us all the answers we need or allow for the recognition and negotiation of the many overlapping and conflicting claims that must be considered in such circumstances. Local factors, in this way, resist "any sweeping fiat from general principles."[82] Certain claims or sets of claims will ultimately have to be prioritized and prevail given their strength and urgency. In some cases, the claims of the introduced animals may ultimately prevail—perhaps the integrity of the ecosystem is not truly threatened. Ecosystems, after all, regularly defy our models and expectations and their resilience may allow new species to be integrated in some way. This may raise challenging questions of domesticity, human interference with nature, and wildness, which also must be addressed. Our answers to these questions will certainly guide future decisions, but they cannot be expected to provide one-size-fits-all solutions. In other cases, if ecosystems are indeed threatened to the extent that they will be unable to support life, difficult decisions involving sentient beings could be required. Desperate times may call for desperate measures, including real evils. This does not necessarily mean that ecological claims are always prioritized over social claims.[83] In fact, what may occur in such circumstances is the prioritization of *certain* social claims—for instance, those of the many other animals supported by an ecosystem, which often *includes* humans—over the social claims of others.

Against minimalism

Though there are real challenges in bringing together the claims of individuals and collectives, Midgley argues that we are more than capable of attuning ourselves to each set of claims without exaggerating one over the other. The animal-environment dynamic, moreover, is probably not the clash with which we should be most concerned. More pressing are the practical issues close to home that do not involve choices between animals and the environment. They are the ones that involve callous and negligent treatment of animals. Animal cruelty and exploitation are problems that we can influence in our everyday lives—in what we eat, which products we buy, and which industries and practices we support. We can, that is, take much more care about the choices we make in light of the many members of our community, human and otherwise. The point of the mixed community, she argues, is that "these creatures are important . . . forming part of a whole where all of them are necessary."[84] Here, we have the ability, in our everyday lives and choices, to transcend the "minimalist" view of morality, which seeks to restrict, rather than widen, our moral horizons.

5

Our Connection to Nature

We rejoice in the whole of nature and being part of it,
as something to which we belong.[1]

Midgley's environmental thought, much like her animal thought, stems from her conception of the continuity of humans and nature. Accordingly, her environmental critique challenges the barriers that have been placed between humans and the natural world. Her targets are usually ideas and attitudes of human separation from and superiority over nature. These anthropocentric (human-centered) attitudes tend to abstract humans from the natural world—for religious and philosophical reasons—and marginalize the natural world as the mere stage for the human drama. This abstractionism, combined with mechanistic views of the world as passive and inert, has long justified careless and exploitative treatment of nature. This treatment has been further legitimated by hyper-individualist ideals of freedom, which frame the world as a resource to serve human interests in the competitive pursuit of goods. Midgley's environmental thought pushes back against these myths—imaginative visions of the world—in favor of more relational and holistic visions of our connection to nature. This vision offers more comprehensive and cooperative ways of understanding our interdependence and interconnection with each other and the wider world. The earth, in this vision, is not an enemy or a resource. It is our home.

Convergent critiques

Midgley's environmental critique converges with her animal critique, spotlighting the pervasiveness of notions of human separation from and superiority over the natural world. Though there are many contributors to animal and environmental ideas and attitudes—acknowledging, for instance, material factors and the mutual shaping influence of ideas *and* practice—these perspectives, to some extent, have parallel roots, at least in "Western" religion and philosophy.[2] This certainly does not mean that all religious and philosophical thought has been antagonistic toward the natural world. There are vibrant traditions of fellowship, care, and reverence for nature. Strains of Greek, Roman, and pagan religion and philosophy venerated nature, often viewing the earth as a life-giving parent.[3] Likewise, Abrahamic theology has, in varying degrees, emphasized care and stewardship toward all of God's creation. Some, for instance, have interpreted Gen. 2:15—"The Lord God took the man and put him in the Garden of Eden to till it and keep it"—as encouraging agrarian ideals of nurturing and cultivating nature.[4] Christian theology, which has perhaps most strongly influenced environmental ideas and attitudes, includes "green" traditions, most notably that of St. Francis, who emphasized communion with nature as God's special creation.[5] These and recent green approaches have engendered lively strains of environmental care in diverse traditions today.[6]

Yet, though green or eco-friendly trends are more common today, they have not been the historical norm. The history of religious ideas, specifically within the Christian tradition, displays ambivalence and real discomfort with the material world of bodies and nonhuman nature. Though Christian thought has often emphasized communion and fellowship with nature, it has also regularly distanced humans, who possess souls, from the material world in favor of the spiritual one. Various Christian myths and theologies have, accordingly, envisioned earth as *below* heaven—the ultimate spiritual world—invoking the value hierarchy of the Great Chain of Being (see "Animals"). The earth below is a symbol of darkness while the sky above is the symbol or source of light.[7] Heaven above is the goal of life while the earth below is a temporary stop, a distracting or even dangerous world that one's soul or spirit is passing through on the way to its final destination. Augustine, for instance, spoke of the inherent battle between flesh and spirit and sought refuge in the only true source of joy and happiness, the world of spirit.[8] The righteous path leads upward, toward heaven, and the degraded path leads downward, toward oblivion or hell. These visions of the separation and subordination of nature and the material world, combined with the Genesis mandate to control and conquer nature, have justified human exceptionalism and dominion (see "Animals").[9] The physical world can then easily be seen as "raw material" or "simply 'stuff' there for us to exploit."[10] Though there are always divergent streams of interpretation—for instance, love of the Promised Land or communion with animals[11]—interpretations that favor the superiority of the human spiritual world have often prevailed.

Midgley's ecological critique is not an attack on Christian theology or doctrines, but rather an illustration of how powerful religious ideas can become integrated into philosophy. Though Renaissance humanists and Enlightenment rationalists considered themselves to have forced out the influence of religious tradition and myths, this is far from the case. Christian environmental ideas were, for the most part, taken as given. Humanist and Enlightenment reformers borrowed, uncritically, the notion that humans are inherently valuable and nonhuman nature is not. The nonhuman world became the mere background for the human drama, a resource to be utilized for bettering the human sphere. This anthropocentric approach was fueled, in large part, by the Cartesian worldview, which dualistically separated mind and matter into distinct substances (see "Human Nature"). The Christian tradition, which itself promulgated a Neoplatonic spirit-matter dualism, engendered and synergized with this Cartesian approach. In both scenarios, nature is squarely in the material realm. Nonhuman nature, in turn, was seen in increasingly mechanistic terms, devoid of the spirit or rationality that characterized morally relevant beings (see "Animals"). The ancient myth of a living, vital, active earth—prevalent in pre-Christian sources—was replaced, Midgley argues, by a myth that envisioned the earth as "mere, lifeless, valueless matter, a dead world of objects without subjects, fit only to be appropriated by us."[12] This myth fueled the seventeenth-century attempt to demystify and de-personify nature, once and for all, and remains influential to this day.

Shifting concepts and horizons

Midgley argues that we live in a much larger drama than has been represented in these anthropocentric visions of the nonhuman world. We must look beyond the narrow scope of this dominant moral landscape if we are to take account of what matters in this world. And this is not just for the sake of the nonhuman world. Human interests are likewise significant considerations in tackling environmental problems. The growing list of global environmental problems that we now face—nuclear waste, polluted air and water, deforestation, climate change, and so on—is not just a set of problems for the nonhuman world.[13] They are *dual* threats to both humans and nonhumans alike. "No environment means no people," she writes, and "a dismal, distorted environment means dismal, distorted people."[14]

Environmental harms and justice

It makes sense that concern for human welfare should include concern for the environment. This is especially important when we consider that the most vulnerable humans are often those that face the greatest ecological threats. Environmental justice demands that we take the claims of those disproportionately affected by climate change and pollution seriously. Small island nations losing coastlines to rising sea levels and African American communities living with higher rates of lead poisoning are prominent examples of the disproportionate threats and harms that stem from environmental problems.[15] Climate change, furthermore, has the added danger of intensifying existing disparities and conflicts around the world while drastically altering the already wildly inequitable landscape of human civilization.[16] For these reasons, a prudential approach to human welfare and the environment is well-warranted—by taking care of the environment, we take care of ourselves, especially the most vulnerable and nondominant. Polluting a river is not good for the river, its various inhabitants, or the humans that drink from and irrigate their crops with it. Likewise acid rain, climate change, ozone depletion, and virtually all environmental harms are, simply put, bad for humans and nonhumans alike. In this way, consideration of human interests is essential and inextricably part of environmental ethics.

Human interests and the limits of enlightened egoism

Yet, there are some limits if environmental practices are only approached in terms of human interests. These interests, Midgley argues, must be balanced with concern and care for nature itself. This is due to some limits in prudential self-interest, which is a version of enlightened self-interest or egoism: the idea that we help ourselves by helping others (human and nonhuman). Hobbes thought prudential self-interest was one of the most powerful motives in human life—identify a threat to our survival and we will make an effort to avert or abate it.[17] This prudential approach animates some current approaches to environmental sustainability, which argue that we must change our practices if we want to sustain our lifestyles and well-being into the future, or, more direly, survive. Midgley argues that this approach can make sense, to an extent. We are typically pretty good at looking out for ourselves ("ourselves" including our species in this case). Yet, for all our self-interest, we are often far less prudent than Hobbes and others have presumed. We regularly engage in willfully self-destructive behavior and ignore dangers to our health and well-being. We can, moreover, be quite narrow in our self-interest, which often blocks long-term considerations on the whole, and especially consideration of others. In other words, our egos can be "rather *un*enlightened and hard to enlighten."[18] If we are to rely on self-interest for long-term environmental change, we face the challenge of competing with this shortsightedness.

From an environmental perspective, the biggest challenge with enlightened self-interest is not just that it appeals to narrow individual interests, but that it is based too heavily on weighted human interests (especially wealthy and elite human interests). It is true that we almost certainly benefit when we protect and restore nature, try as political and industrial interests might to pit human interests against environmental ones. And to a certain extent, *any* environmental care is a step in the right direction, whatever the motivation. Midgley's critique, however, is that enlightened egoism only gets us so far. As in the human realm, egoism perpetuates the notion that we should only benefit others insofar as we ourselves are benefited: "What's in it for me?" This mentality furthers a competitive vision of the world that only encourages concern for others when it is in one's own interests (see "Morality"). Too often, this boils down to a preoccupation with economic interests to the neglect of all others. "We can see people," she writes, "busily engaged in sawing off the branches on which they (along with many others) are sitting, intent only on getting those branches to market before the price of timber falls. Prudence does not prevent this destruction."[19]

Prudential self-interest, in this way, has a role, but has some inherent limitations. Such concepts can only be seen as aspects and temporary measures in the more comprehensive movement to change our moral landscape. Midgley argues that wider awareness of the nonhuman world demands "some kind of adaptation from us, adaptation of a morality that was formed for a quite different, more manageable kind of world. We can't go on acting as if we were still in that world. On that path, there is no way through."[20]

Paths forward

Fortunately, there are paths forward in expanding our moral horizons. There are many ways of recognizing interests beyond our own, including those of nonhuman nature and even the biosphere as a whole.[21] Our priority systems can bring together a range of concepts rather than a simple "conceptual monoculture," providing us with a variety of tools for weighing the claims of the nonhuman world.[22] "When we ask," she writes, "whether principles such as respect for life, stewardship and species' rights can be used, the answer is surely, Yes—and plenty more. All have their own advantages and drawbacks. All will need further development. . . . All overlap on an enormous area of the action that is necessary. All, however, suit some contexts better than others."[23] There is, in this more pluralistic and pragmatic approach, potential efficacy in appealing to prudential human interests. Preserving a rain forest really is good for everyone involved. Yet, on the other hand, we can also see the rain forest as important in itself—as valuable in its own right, non-instrumentally. Whether it be because of the characteristics, capacities, and/or purposes of organisms (individualism), or because of the emergent value of collectives, systems, and the whole (holism), philosophers have compellingly argued that there is value in nonhuman organisms and entities (see "Animals").[24]

Midgley supports this appeal to the value of nonhuman nature itself, though she cautions against stretching such terms as "value" too far beyond their everyday sense, lest they fall victim to some of the technical pitfalls of other categories of moral worth (see "Animals"). Valuing nature intrinsically has an important role to play in our moral deliberations, but she sees this insight as a necessary aspect, again, of a more comprehensive value system that takes account of *many* things that are important in the world. This multiplicity is what our moral visions should imaginatively describe and illuminate. The challenge with nature in these visions is that we are grappling with how to articulate such immense ideas, reflecting the "open-ended problem" of "what attitude creatures like ourselves ought to take toward the cosmos within which we are so small a part."[25]

Historically, this imaginative process, informed by the aforementioned religious and philosophical ideas, has consistently nurtured human abstractionism and exceptionalism. Changing this untenably anthropocentric vision of the world involves, on the negative side, removing the conceptual barriers that have been placed between humans and the natural world and, on the positive side, shifting toward more harmonious visions of living with nature. For Midgley, both the positive and the negative approaches, at a base level, involve developing and encouraging more relational, cooperative, and interdependent ways of seeing ourselves. The obsession with individual freedom, though historically warranted, can hypertrophy and become more harmful than helpful, isolating and separating the self from others and the natural world while increasingly narrowing our care and concern (see "Human Nature" and "Morality"). This hyper-individualism tends, again, to not only overlook our relations and interdependencies with others, but also slices up the world into competing units. Others are seen as competitors for limited resources rather than cooperative community members working toward the greater good. This mentality exaggerates freedom *from* (relations, bonds, responsibilities) while neglecting what we are meant to be free *for* (caring for others, visions of common good).

Midgley argues that notions of individual freedom that neglect our relations and interdependencies can encourage a callous and even hostile view of others, including the natural world. Freedom *from* nature can very easily transform into freedom *for* exploiting it. Thus, the same ideals of freedom that can warp our conceptions of the self, morality, animals, and gender also contribute to our narrow visions of the natural world. While it is essential that individuals understand themselves as free—as capable of effort and self-direction in the world—we must do better in also recognizing and valuing the relations and dependencies that enable our freedom and, to an extent, constitute who we are as whole persons. This must include more positive visions of our connection to nature.[26]

Our connection to nature

Midgley envisions our connections to each other and the natural world in terms of "wholeness and separateness."[27] We exist as individuals, but we are parts of wider collectives of varying types and intensities—families, communities, ecosystems, and so on. As individuals we are deeply entangled with each other and these overlapping collectives.[28] Utilizing the image of a tree, she writes of the "variety of asymmetrical relations found within a whole. Leaves relate not only to other leaves, but to fruit, twigs, branches, and the whole tree. People appear not only as individuals, but as members of their groups, families, tribes, species, ecosystems and biosphere, and have moral relations, as parts, to these various wholes."[29] She reminds us:

> Of course, human beings are distinct individuals. But they are also tiny, integral parts of this planet—framed by it, owing everything

to it, and adapted to a certain place among its creatures. Each can indeed change its life, but does not originally invent it. Each receives life in a family (as a petal does in a flower), in a country (as the flower does on the tree), and in the biosphere (as the tree does in the forest). Our environment gives us nearly everything we have.[30]

Our environments here are not alien entities or mere aggregates of competitors, but the vibrant context of the self, a self which "unavoidably looks for its fulfilment to horizons far beyond its private destiny."[31] This more relational and holistic way of thinking may not always be forefront in our visions. Yet, it is essential if we wish to alter our treatment of the nonhuman world. This, as Midgley writes, is because our visions shape our attitudes and priorities, and are "crucial for our moral attitude. When this larger imaginative vision changes, the light in which we see all our various concerns is altered. Priorities shift, carrying a corresponding change in duties."[32]

People can and often do look to this wider horizon. In spite of the preoccupation with freedom and individualism, Midgley argues that the burden of proof should not be on those who emphasize and appreciate our connection to nature, but on those who wish to destroy it.[33] As part of this wider community of life, the fates of other beings need not be a matter of indifference to us. In fact, we often feel and express this connection to the greater whole. Midgley expands:

You feel akin to the whole thing. And that, I think, is probably a central point. We rejoice in the whole of nature and being part of it, as something to which we belong. It follows that if some important part of it is threatened, then we take alarm, as if for ourselves, so to speak, but presumably in proportion. Nature is a whole of which we are quite a small part. . . . If we hear news of the destruction of a forest, the point is it is not something totally alien to us. It's not something to throw away like last month's newspapers. It *concerns* us. It's how we identify ourselves, isn't it? What we feel ourselves to be. People, surely, mostly, have thought of themselves as a small part of something much larger.[34]

Holistic visions

Given this relational and holistic emphasis, Midgley gravitates toward more ecosystemic ways of envisioning nature. These approaches, which focus on the interconnection and interdependence of organisms, positive feedback mechanisms, and local and global patterns of natural systems, tend to go much farther in capturing the dynamism and vitality of the natural world and its entangled constituents.[35] And, significantly, rather than treating wholes and collectives as mere aggregates of individuals, holistic and ecosystemic approaches recognize that "wholes and parts are equally real," and that each is unintelligible without the other.[36] One such vision that integrates ecosystemic thinking in the largest scale is the notion of Gaia. Originally advanced by James Lovelock, Gaia is a scientific vision that sees "Earth and the life on it as an active, self-maintaining whole."[37] Gaian theory, which draws its name from the ancient Greek earth goddess, maintains that this great whole is constituted by the totality of systems and their organic and inorganic parts, all of which are connected to all others through complex webs of interactions and dependencies. From local interactions between organisms and their environments to global weather patterns and currents (such as the massive flows of Saharan dust to the Amazon basin and the Amazon's "Rivers in the Sky"), the earth can seem like an active organism. Life, in this picture—shown most vividly in satellite imagery—is not simply "a loose, chance jumble of competing entities but an interdependent system, a symbiotic whole that keeps itself going by a constant interchange of benefits between its parts."[38]

This vision is by no means new. Plato and the Ancient Greeks often considered the earth to be one single organism.[39] However, this way of thinking fell out of favor when Christian leaders deemed it as heretical and pagan. While the earth may have had certain value as a creation of God, it was still a created and subordinate object and not itself sacred. Later, mechanistic science, under the influence of Descartes, categorized and studied nature (matter) as passive and inert rather than active and dynamic. In this way, the natural world was increasingly objectified and distanced from its ancient dynamism. Yet, Midgley argues, twentieth-century paradigm changes in physics and systems ecology have challenged this vision of the world. Quantum mechanics has undermined the notion of an orderly mechanical world composed of distinct, disconnected, and predictable atoms in favor of a more dynamic, interrelated, and indeterminate notion of the physical world. Ecology and chaos theory, furthermore, have drawn attention to the dynamisms of global systems and the webs of interdependencies that characterize all aspects of the natural world. The world, from the smallest to the largest scale of measurement, appears intrinsically related and interdependent.

It is tempting, Midgley notes, to exaggerate the notion of Gaia—a scientific hypothesis and vision—in anthropomorphic and even religious ways. This is a quite natural response for some: Gaia has large-scale mythic elements that are all the more dramatized in the midst of a global ecological crisis. And, given its symbolic elements and original connection with Greek religion, it, like other scientific visions, has the potential to verge into the religious (see "Religion and Science"). She muses that these religious elements, plus personification of the Gaia concept—not to mention its gendered elements (see "Gender")—have led some to see Gaia as a "new age fancy" instead of a useful *metaphor* for envisioning global connectivity and interdependence.[40]

Midgley employs the concept of Gaia as a *metaphor* for understanding the interconnections of the earth as a whole. She argues that Gaian thinking, whether by that name or others—for instance, Lovelock's more recent "medical model"[41]—is one way among many, including atomistic and mechanistic models, of picturing the world. These models and ways of seeing the world are chosen and evaluated through many factors. (Religion, science, gender, and so on likely factor in here, much as they do in our visions of human nature, the self, and so on.) Since our visions influence our actions, Gaia and other holistic ways of thinking might help us in understanding our role and *impact* on the earth. The whole system, in these approaches, is vulnerable. This vulnerability does not mean that the system can be destroyed, but rather that it can, in its current state, be damaged or altered in response to stimuli (i.e., human intervention and harm). And, on a more ominous note, will carry on in whatever state, with or without *us*. Thinking in these more global, interdependent terms is a direct counter to the hubristic view that humans are at the center of the cosmos. The earth, in this more humble holistic picture, is not here to serve our every need.

Holistic thinking, in this way, might help us out of some of our more abstractionist and exploitative visions. "I regard [Gaia]," Midgley reiterates, "as a *myth* about the earth, and a good myth, that the Greeks and many others have had of the earth as a parental relation to us. This means that we *owe* it a great deal. . . . It's not just something that we have been given to work with."[42] This relational view of the earth, she emphasizes, is "pretty important because people really do need this sort of inclusive place within which everything takes place. We really do need the unity."[43] There are, of course, a number of ways to conceive of this wholeness and connectivity, but the image of Gaia is something of a ready-made shorthand for this worldview.[44] Regardless of the exact language or metaphors used, Midgley argues that these holistic concepts better capture the collectives or wholes in which we live, as well as our interdependence with them. Nature is not a "static background," but the whole of which we are part. Seeing ourselves as part of this larger community is not fantasy, but a reasonable understanding of the earth's systems and inhabitants.

Relationality and holistic claims

More holistic visions can, furthermore, provide us with a framework through which we can take ecological claims more seriously, something that individualist accounts neglect, struggle with, or are incapable of doing. Since the early days of animal and environmental ethics, reformers have sought to extend individualist visions and frameworks to animals (see "Animals"). This has been highly effective in certain cases and has brought about much practical change in the area of animal liberation in particular (Singer, Regan, etc.). However, when these sentience-based approaches are extended to cover the wider world—plants, trees, species, non-sentient animals, rivers, ecosystems—one faces the challenge of applying individualist concepts to non-sentient beings and entities.[45] Some theorists have, as such, made forceful cases for the value and purposes of individual non-sentient organisms.[46] This is certainly a significant part of the equation. However, for nonliving and collective entities, Midgley argues that, in isolation, "individualism is bankrupt of suggestions for dealing with these non-human entities."[47] In other words, individualist categories and concepts such as rights or interests, as useful as they may be, are not exhaustive in their application to the nonhuman world. More is needed.

Midgley's discussion of social and ecological claims is again helpful in approaching the nonhuman world (see "Animals"). Duties and obligations to non-sentient beings and collectives (ecological claims) are of a different sort than those to individual sentient animals (social claims).[48] This does not mean that either set of claims has an inherently superior status. Rather, they are just of a different kind. And, again, they are variable, sometimes overlapping and continuous, sometimes conflicting. Midgley resists the temptation here to elevate or overemphasize collectives or wholes. Though she staunchly resists hyper-individualism, she also pushes back against unrealistic exaggerations of the wider whole: hyper-holism. Gaian advocates, deep ecologists, and other holists can, at times, make it seem as though we are *identical with* or *subservient to* ecological collectives or the biosphere as a whole. It is, Midgley counters, possible to create a balanced vision of organisms and the world without losing the individual or the respective collective(s) in which they are embedded. "Concern for the whole," she writes, "and concern for individuals are simply not alternatives. They are complementary, indeed inseparable, aspects of a decent moral problem."[49] Furthermore, "neither of them is reducible to the other. It is always possible for the two to conflict, but it is always necessary to try to bring them into harmony."[50]

Ultimately, reconciling both sets of claims—individual and collective, social and ecological—is important and largely possible (see "Animals"). Midgley argues that there is nothing limiting us from this type of wider concern and care for diverse *types* of claims. This stretches beyond the realm of sentience—where individualist approaches often meet their terminus—pointing toward our real ability "to care for such things as trees and forests, grasses, rivers and mountains."[51] Claims of species and rivers matter to us, even if these entities are nonverbal, non-sentient, or otherwise lack the capacities that may classify one as having "worth" in certain approaches. Our special moral capacities, she writes, enable us to "perceive and respond to consciousness in others in a special way. But as beings forming a small part of the fauna of this planet, we also exist in relation to that whole, and its fate cannot be a matter of moral indifference to us."[52]

Wider horizons

Midgley argues that this "care for the rest of the planet, not only for ourselves," is one of "the special qualities that make humanity worth preserving."[53] Our expansive sense of care and concern, if cultivated and allowed to flourish, is among the true graces of our species. Now that we are fully aware of our ability to influence the planet, what will we do with it? Will this power be used for good or ill? How will we treat *our home*? Midgley proposes that a whole range of values and visions are available to us for correcting and balancing unrealistic anthropocentric myths.

There are, to be sure, many other contributing factors to our wide-reaching ecological problems, most of which cannot be taken up in this volume—political and economic factors, patterns of consumption, and so on, most of which connect up with pernicious environmental ideas and attitudes. We can, though, highlight the problems in our narrow visions even further by drawing attention to the influence of *andro*centrism, or male-centric, worldviews. Ecofeminists and others have long pointed out that this posture of dominance has been especially damaging in our conceptions and treatment of the natural world.[54] Male-centric individualism, moreover, has had sweeping consequences in relation to virtually *all* of the issues covered so far in this book. The heroic hypermasculine ideal of freedom, in particular, is one of Midgley's main targets. I take this up, along with connected issues, in the following chapter.

6

Gender and Fragmentation

Because people weren't talking about gender, they weren't aware of how many different things you might say about it.[1]

Mary Midgley's philosophical career spanned all four "waves" of feminism. As such, she engaged a range of issues related to feminism, sex, and gender, though these topics were rarely her main focus. In the 1980s–90s, Midgley and her regular collaborator, Judith Hughes, drew attention to the some of the connections between gender and wider problems in the Western philosophical tradition. One of their more prominent criticisms centered on the masculinist ideal of freedom, which tended to elevate (presumed) male virtues while perpetuating false hierarchical antitheses between men and women: reason versus emotion, mind versus body, masculine versus feminine, and so on. Midgley argues—with Hughes and elsewhere—that combating masculinism and this oppositional and competitive way of seeing the world should be among the main priorities for feminism and philosophy writ large. Masculinist individualism fragments our notions of the self, dividing *aspects* of ourselves against each other while abstracting and isolating us from others and the world. Midgley and Hughes, in turn, suggest pragmatic changes and countermeasures for both the short term and the long term (including, for instance, that "work" be more responsive and accommodating to women) in combating this fragmenting vision. They also argue that progressive, wide-scale change in gender-related issues involves addressing a range of interconnected problems and injustices that plague the world. Though the landscape of feminism and philosophy has dramatically shifted since the 1980s–90s, Midgley's insights, along with Hughes, offered a much-needed critique of the gendered aspects and practice of philosophy.

A legacy of negligence

Midgley's gender critique goes straight at the core of Western philosophy, which is riddled with problems. These problems are typically "open and flagrant," rather than elusive or covered up, making them easy to spot.[2] They range from negligence on the topic of gender to outright hostility toward women. For all the intellectual inheritance and significance of the Western tradition, we are left with a confused muddle of inconsistent and incoherent ideas on gender and, especially, on women. Women's perspectives are rarely taken into consideration and, when they are, it is often in ridicule or in contrast with thinking, rational men. This rightly perplexes feminist philosophers, Midgley observes, because the topic of gender is where intellectual rigor tended to meet its demise. "It is not just that some of what they say about women is false," she writes, "nor even that some of it is wicked. What startles is its silliness. When this topic is raised, normal standards of thought collapse."[3]

Aristotle wrote that "the courage of a man lies in commanding, a woman's lies in obeying" and that "the female state" is "as it were a deformity," and Hegel thought that "men were like animals, women were like plants—that is, without any real individuality."[4] Though there are certainly attempts—for instance, John Stuart Mill's critique of gender inequality[5]—that openly and critically address gender issues, other historical treatments fall into the categories of absurdity or animosity. Schopenhauer, for instance, made no real attempt to hide his disdain for women:

Hers is reason of very narrow limitations. This is why women remain children all their lives, for they always see only what is near at hand, cling to the present, take the appearance of a thing for reality, and prefer trifling matters to the most important. It is by virtue of man's reasoning powers that he does not live in the present only, like the brute, but observes and ponders over the past and future; and from this spring discretion, care, and that anxiety which we so frequently notice in people.[6]

Later, Nietzsche wrote that "man should be trained for war and woman for the recreation of the warrior; all else is folly."[7] This leads Midgley and Hughes to observe that "there cannot be many matters on which Freud, Nietzsche, Rousseau and Schopenhauer agree cordially both with each other and with Aristotle, St Paul and St Thomas Aquinas, but their views on women are extremely close."[8] Do these figures, and others like them, dodge some criticism by being products of their times? Is it anachronistic to judge them by present-day standards and sensibilities? Midgley and Hughes think not, as all major philosophers had predecessors and contemporaries who approached gender in better ways. (They mention Euripedes and Mill, for instance.) Women reading philosophy, they note, should have no embarrassment in criticizing "the ludicrous views" that stem from their negligence.[9]

There are surely numerous social and personal factors involved in this strange history. Power dynamics and the reluctance of male elites to relinquish power certainly play a major role. Midgley also observes that many of the most prominent Western philosophers were bachelors or celibates, an influence from the real world which perhaps influenced their thought.[10] However, there are other issues involved, only some of which can be addressed here. Midgley argues that many of these problems stem from the same damaging ways of seeing that world that distort most of the topics addressed in this book. Dualistic thinking, for instance, in its philosophical and religious forms, creates a whole network of confused and unquestioned antitheses. Aspects of life are conceived of as warring alternatives or enemies instead of complementary parts of the whole. Male and female are envisioned as binary opposites and grouped with a whole range of other antitheses, including reason versus emotion, mind versus body, independence versus relationships, and science versus nature. These oppositions not only illicitly separate complementary elements of our lives and fragment the self, but they also tend to elevate one side of the antitheses—the collectively masculine side of the opposition—above the other. Moreover, the other side of the division that which is to be ruled.

Women have consistently been portrayed as the antithesis of the male ideal. Aristotle, again, equated the identity of women with their function, which he identifies as sexual and supportive to men.[11] This functional view of women has, to one extent or another, remained influential to this day. The function of men is rational thought and contemplation—the domain of philosophy—while the function of women is bodily and reproductive—the domain of the domestic. It is for this reason that Kant and others chastised women who tried to "take on" the active roles of men.[12] Men actively change the world while women watch and support them. Women are, furthermore, inferior in this simply because they are *not* men. They are lacking. They are portrayed as a passive, homogenous grouping against the variegated, active grouping of male agents. In this categorization, all women have the same characteristics, desires, emotions, functions, and so on.[13] This convenient lumping and separating of women clearly should have been criticized by subsequent philosophers many times over. Yet, this negligent approach was too often absorbed and perpetuated by later thinkers, mostly because, Midgley and Hughes write, "the vested interest was, until recently, too strong for any real attention to be possible."[14]

A more modern revision of this thinking resides in the *individualist* male ideal. This vision, for Midgley, is perhaps the most significant contributor to more recent philosophical problems in sex and gender, much as it is with many problems in this book. The Enlightenment ideal of the isolated will, characterized and guided by reason, developed over and against emotions, the body, and relationships.[15] A "rational individual," following Descartes's imagery, is here conceived of as unbound by these limitations on freedom and, moreover, as masculine. Women were characterized both as *impediments* to male freedom—something to be freed from—and *means* to male freedom. That is, as obstacles *and* assistants in the male pursuit of freedom. In both cases, they are not conceived of as free individuals. Women *have* destinies, which are typically sexual and domestic— as members of Hobbes's patriarchal "little monarchy"—while men *choose* their more heroic and transcendent destinies. "A man,"

Midgley and Hughes write, "is entitled to behave as though he were in some sense a shaper of that destiny even if his belief is mere illusion. In general, it is assumed that men choose; women are chosen."[16] An "alien anthropologist," Midgley muses, "would conclude fairly confidently that women were a small minority—a marginal, outlying, anomalous group . . . with some peculiarities which called for special social arrangements, but so untypical as not to affect the general shape of society."[17] This, of course, is not to diminish the many significant contributions of women to society, but rather to highlight the absurdity of these prevailing ideas and attitudes.

Interconnected antitheses

These wide-ranging ideas and antitheses had expansive ripple effects as they spilled over into many areas of thought and life. In the history of science, this is seen in the project of demythologizing and desacralizing nature. Of the many guiding images that could have been cultivated by the architects of the Scientific Revolution, combative gendered imagery rose to the top: nature, often referred to in gendered terms, was a foe to be subdued. This was mainly due to the influence of the conceptual connections on the left side of the prevailing antitheses: male versus female, reason versus emotion, science versus nature, true religion versus paganism, and so on. Midgley argues that these illicit hierarchical antitheses have held thought captive. They reinforce visions of the world in competitive, mechanistic, and militaristic terms, elevating masculine virtues and marginalizing ones that were considered feminine, such as cooperation, gentleness, and nurture. During the Scientific Revolution, the female principle (in Robert Hooke and others) represented life and bounty but also an underlying sense of mystery and darkness.[18] The sacred elements of nature, with roots in pagan religions, often carried more feminine symbolism and personification, whether it be as a nurturing mother or sustaining goddess, such as Gaia (see "Nature").[19] The male principle, on the other hand, represented order and the divine light of God, who was unequivocally viewed as male.[20] The demythologization and desacralization of nature thus required the domination and subjection of these female, pagan images (not to mention their adherents) in a crusade of light over darkness. Rather than integrate the male and female principles into more complete visions of humanity and nature, influential voices in the Royal Society, for instance, set them to battle against each other. This quest for a "truly masculine philosophy" was characterized by vivid gendered language evoking the submission, subjection, and penetration of nature.[21]

The point here is not that this imagery of femininity was accurate or preferable. Rather, it is that women and nature, as such, were viscerally portrayed as enemies to be subdued by male science. The "masculine birth of time," as Francis Bacon termed it, promised to free men once and for all from the rule of nature as a demanding mother or "irrational queen."[22] Men, as strong rational beings, needed to rule women, who were weak, irrational, and emotional. Rather than seeing reason and emotion as complementary aspects of *all* human life, they were portrayed as gendered attributes, respectively, symbolizing power and weakness. Midgley argues that this fanatical opposition of male and female principles clearly points toward wider gender dissonance, and that one need not be a "full-time feminist" to recognize these historical problems:

> When a school of thought, officially dedicated to clear, literal, unemotive speech, regularly uses a lurid language of sexual pursuit, torture and rape to describe the interaction between scientists and the natural world, trouble is also surfacing about the relations between actual men and women. At such a point, an entry in the index under the heading "gender insecurity" doesn't seem excessive.[23]

Male domination, however prolific and insidious, is prone to such insecurity. This was especially the case in an era characterized by political revolt and the rise of individualism, which threatened European male power in the world (though, Midgley notes, political and egalitarian reformers show great adeptness at neglecting and dismissing those outside their own narrow circles).[24] The repudiation of femininity—in both the world and men—served only to exacerbate and accentuate the deep tensions and insecurities surrounding masculinism. This distorted and exaggerated maleness harms everyone and everything, and nature is an easy target on which to project and extend these gender conflicts and tensions.[25] Any victory over nature, or discovery of "her inmost secrets," symbolized a victory in the "gender-battle," the language of which still accompanies "scientistic celebrations of human dominion over the earth around us" today.[26]

Nature, inert and passive and removed of "her" sacrality in the name of religion and science, was thus at the mercy of men, much as women were expected to be. This coincided with a much wider set of oppressions in which growing empires viewed all things as rightfully belonging under their domain: women, children, nature, and other territories and cultures. This colonial mindset, much like the wider sets of interconnected antitheses, lumped together numerous hierarchical oppositions.[27] Here, again, the Western tradition, which should have scrutinized and questioned such groupings and prejudices, regularly failed to engage the conceptual problems involved in these assumptions and dynamics. Silence and evasion were the tenor of the age, and examinations of the intersections of male and cultural domination, when they are addressed, are predictably inconsistent and erratic. Midgley writes: "On the rare occasions when these theorists do not run, but consider [women, slaves, and other races] more fully, their discussion often looks as if they had not written it themselves, but had left their paper for the afternoon to some weird secretary who wanted to discredit their doctrines."[28]

This erratic approach to women and other interconnected problems is visible throughout egalitarian literature. Egalitarianism is typically motivated by equality for a select group.[29] Historically, this group was usually composed of male citizens demanding their share of respect or rectification of injustices. These injustices are, to be sure, real and worthy of redress. And, as a tool for accomplishing this task, egalitarian movements have often been successful at just that. Yet, this narrow focus is a prominent example of what Midgley terms "the paradox of one-way equality" in which "inequalities above one's own level tend to be visible: those below it to be hidden."[30] Midgley highlights the "self-righteous revolutionaries who expected their women to type the manifestoes and bring the coffee, but remain otherwise dutifully silent."[31] Rousseau, for instance, wrote that "woman is made specially for man's delight," and Stokely Carmichael famously quipped that "the only place for women in the SNCC is prone."[32] These sentiments express the ever-potent assumption that all men should be considered equal, but not necessarily with women.

Masculinism and feminism

Midgley locates these problems in thinking (or not thinking) about gender within the wider project of negotiating the human condition, a project which has been dominated by negligent and masculinist voices and patterns. It is no surprise, then, that these patterns of thought have deeply fragmented the ways in which we envision human nature and, in turn, gender. These concepts, as Midgley argues, shape how we act and live: the neglect, dominance, and injustice promoted by these positions and their policies have real-world implications and ramifications in our everyday lives. For these reasons and many other reasons, philosophy needs feminism. And, it is no surprise that feminism rose to meet these challenges. Philosophy had long been in need of course correction, and reformers such as Mary Wollstonecraft and Simone de Beauvoir were among the more recognizable figures that took up this challenge and fueled a range of movements that could variously be considered feminist.[33] Feminism itself is a notoriously diverse collection of movements, theories, and reformers, and there are a range of factors that contribute to its success. While it is impossible to characterize feminism as any one thing, Midgley and Hughes endorse the stance that "women suffer systematic injustice because of their sex" is something of a unifying thread in feminism.[34] Feminists, in this sense, work "to redress injustices peculiar to women."[35] This approach, furthermore, is "an element which we all need for our thinking on a great range of important matters—social, political, psychological and moral—and whose absence has always weakened that thinking."[36]

Sex and gender are central concerns within feminism, but Midgley stresses the importance of connecting these emphases with issues related to postcolonialism, race, class, war and peace, animals, the environment, and so on.[37] Much of the development of more recent feminism and ecofeminism has involved grappling with these dynamics, especially in challenging some of underlying assumptions of feminism, assumptions which have, at times, been embedded within comparatively privileged standpoints (e.g., from within more dominant cultures). This issue, in part, is due to the legacy of masculinist conceptions of the world, which largely portrayed "women" as a homogenous essentialist category. Midgley and Hughes resist this abstract "lumping," which implies an assumption that women are a single group or "a standard article facing a standard set of problems."[38] They argue, instead, that "the problems women face in different cultures differ hugely; and since their style of life is anyway different . . . the changes they need are also usually different."[39] As such, "it is perfectly natural to speak of all these women, when they try to lead movements to improve the conditions of the women around them, as feminists."[40] That is, there is no one way to be a feminist.

Resisting masculinism

While Midgley does address elements of more recent feminism, she primarily focuses on issues that are more associated with second-wave feminism. Midgley and Hughes, for instance, examined the tensions that arose in the second-wave emphases on equality and individualism. Socialist feminism emphasized "equality and working towards it through normal political channels," while radical feminism emphasized "liberty from male domination" and, at times, advocated separatism.[41] There were valuable elements in each of these approaches. Socialist feminism drew attention to the economic oppression of women and systemic political injustices. It was especially effective in the public sphere and had strengths in engaging the political elements that required reform. Radical feminism drew attention to the more general oppression inherent in asymmetrical gender relations in patriarchal cultures. It was especially effective in the private sphere—partly through arguing that the personal *is* political—and had strengths in engaging the social and psychological elements of sexism.[42] Each set of problems required drastic change as women were and are oppressed in these ways and more. However, Midgley and Hughes observed that partisans of each approach tended to push for a unanimous movement in which "she who is not with us is against us."[43] Difference of opinion between these (and other) movements was sometimes seen as a betrayal of the cause.

Unanimity, however, is not always possible. There may be general agreement on certain problems, but unity on *the* problem and *the* solution often eludes any movement. Midgley and Hughes argue that one of the points of contestation within feminism—and certain other movements—is most clear in the general tension between equality and liberty.[44] These are both things that we want, but their attendant policies may be quite different. The principle of equality promotes restraint and intervention in an effort to ensure equal treatment, while the principle of liberty, on the other hand, promotes freedom from restraint and intervention. These principles are not necessarily irreconcilable but require a level of compromise in order to coexist in some type of harmony. The challenge that Midgley and Hughes saw is that polarizing approaches to solving women's issues can exacerbate, rather than accommodate, the disconnect between these divergent principles instead of focusing on important overlapping concerns, ideals, and goals.

Egalitarianism and individualism, which, in varying measures, respectively, rely on the principles of equality and liberty, are important aspects of any reform movement. However, each of these approaches to reform has some conceptual baggage and unhelpful elements (See "Animals" and "Nature"). In relation to gender, they have conceptual connections to and tendencies toward the male ideal. Egalitarianism, Midgley argues, tends to homogenize, flattening out the diversity both within and outside of groups.[45] She is concerned that efforts to achieve equality can—consciously or unconsciously—promote a version of masculine "sameness" and the male ideal. Here, rather than valuing masculine *and* feminine qualities (whatever those may look like as parts of the whole person), equality can look like an effort to disown the feminine. Reformers can lean heavily toward the position that there are no relevant differences between men and women, even though acknowledgment of difference can be an important factor in fair and equitable treatment—for instance, in issues of sexual harassment, maternal care, and access to contraceptives. Her point is that equality is a *moral* principle—it does not require sameness.[46] Of course, the nature and scope of sexual differences— which are distinct from the construction of gender—and how much they influence human life is a matter of great debate.[47] Whatever these differences may or may not be, she argues that we must understand rather than downplay or ignore them. This is because, if we want things to change, we must, so to speak, "understand the permanent."[48] Denying the body, furthermore, can perpetuate such masculinist antitheses as man versus woman and mind versus body. For Midgley, biology does not equal destiny—one might imagine for men, women, or any other gender identity—and difference does not equal inferiority. The problem, rather, resides in the value hierarchies, prejudices, and conceptual baggage that have been built into any differences. In this way, feminist opposition to talk of difference is well-warranted, given how the concept has been abused.

Individualist movements, on the other hand, can likewise lean too heavily toward the male ideal. Ideals of freedom tend to center on the notion of freedom *from*, more than freedom *for*. Traditionally, the collection of items to be free from included, among other things, bonds and relationships, along with emotions and the body. These were, accordingly, associated with women. Women were encumbrances for men to be freed from while they were, at the same time, the invisible support for men. Thomas Carlyle, for instance, said of his wife: "How loyally and lovingly she had made herself a soft cushion to protect him from the rude collisions of the world."[49] Midgley summarizes this conundrum:

> The whole idea of a free independent, enquiring, choosing individual, an idea central to European thought, has always been essentially the idea of a male. It was so developed by the Greeks, and still more by the great libertarian movements of the 18th century. In spite of its force and nobility, it contains a deep strain of falsity, not just because the reasons why it was not applied to one half of the human race were not honestly looked at, but because the supposed independence of the male was itself false. It was parasitical, taking for granted the love and service of nonautonomous females. . . . It pretended to be universal when it was not.[50]

This parasitic and hypocritical vision of freedom permeates individualism and can encourage abstraction from the body, emotions, and relationships in the tradition of the male ideal. Midgley would rather reevaluate the whole set of ideals and virtues, emphasizing the role of care, concern for others, and cooperation alongside the merits of individualism: "This equivocal, unrealistic attitude to the mutual dependence central to human life does not just inconvenience women. It falsifies the whole basis of life. Morality becomes a lop-sided melodrama. The virtues and qualities we need for love and service are uncritically despised, while those involved in self-assertion are uncritically exalted."[51]

The pragmatics of change

These challenges are not insurmountable, especially given cooperation. However, Midgley and Hughes are concerned that the warranted responses to the issues surrounding patriarchy and sexism have resulted in exclusivist extremes. That is, they are posed, again, in antithetical terms. Rather than being framed as complementary approaches that deal with aspects of the problem as a whole, the positions are opposed against each other as alternatives. This coincides with wider sets of false dilemmas that can stem from singular outlooks:

> Will you have equality or liberty? Peace or justice? Employment or leisure? Feeling or reason? Bath or breakfast? Children or a career? Is society formed by the social contract or by natural feeling? Are women just like men or quite different? If you try to say "both" you are accused of wanting to eat your cake and have it, and it is added that she who is not with us is against us.[52]

Like religions, reform movements have great power and promise in shaping the world. However, the more movements look like religions—in all their passion and motivational force—the more they can take on the less desirable aspects of religion, such as the tendency toward insider versus outsider ways of seeing the world. This can inhibit long-term change, which requires that groups work together toward a better world, however difficult that may be.[53] Large-scale social, political, and moral transformations may require many approaches and many solutions.

Moral issues are typically too complicated, variegated, and challenging for any one approach to have all the answers (see "Morality"). Midgley's stance is that "all evils are complex and need to be attacked from many sides at once."[54] There are whole sets of reasons why domestic abuse and sexual harassment are wrong and, accordingly, whole sets of potential solutions.[55] The principles involved in tackling these problems may often diverge, but this does not mean they cannot be more or less reconciled. Moreover, complete reconciliation is often less important than achieving commonly shared goals. Diverse people and principles can work together in opposing gross injustices for differing reasons. This is quite often the solution in the short term, where diverse people and groups work together on particular causes. From there, longer-term social and political changes may be pursued. To return to the gardening metaphor, many hands are needed to weed the garden (see "Morality"). Likewise, many seeds of change—in all their varietals—are needed.

Change and direction

Midgley and Hughes thus present a moderate and pluralistic approach to practical reform. They caution against separatist and revolutionary tactics, arguing they are unlikely to be successful in instigating sustainable change.[56] In the short term, they encourage policies that take a "woman-shaped life" seriously, whatever that may look like.[57] This is because "human life" has too often been assumed to be *male life*. They argue that it is possible to look at women's lives and experiences in ways that do not perpetuate masculinist assumptions. As such, they advocate re-envisioning both men's and women's lives apart from the abstract and damaging assumptions of the male ideal.

This re-envisioning should also involve reexamination of work. Midgley and Hughes argue that modern industrialist capitalism is hardly good for anyone and "suits many men as well as women very badly."[58] This system—which favors production and consumption, merit and reward—unrealistically glorifies competition and "workaholism" over other visions of the good life (including more communal and cooperative visions). Here, success is defined by wealth, hours worked, and goods consumed. Midgley and Hughes question whether women (or anyone) should pursue an equal share of this type of life. "Is it a good thing," they write, "to have hardly any time to see one's children? Or to have one's value measured by one's wage packet? Or to overwork so that one dies of thrombosis at fifty? Unless these are good things, it is not obvious why women should demand an equal share of them."[59]

Midgley and Hughes argue that change should come through augmenting what is expected of *everyone*. Too often the choices presented to women are limited and extreme. "In general," Midgley and Hughes write, "both from feminist and anti-feminist writers, women still get a fairly sharp ultimatum: either (1) have kids and give up your life for them, declining into an irreversible domestic Nirvana which (according to taste) can count either as death or fulfillment; or (2) fulfil yourself and have no children. . . . Do not attempt both assignments."[60] While men have many choices regarding work, family, love, and sex, women have two straightforward and opposed options. Midgley and Hughes encourage the possibility of working out more realistic possibilities between the alternatives, which includes broad changes in the expectations of men *and* women.

Better worlds

Much has changed in the world, feminism, and philosophy since the 1980s. Some of the recommendations that Midgley and Hughes advocated, such as the expectation of shared parental and domestic duties, have become more common (at least in some places). Some of the issues that Midgley and Hughes addressed, on the other hand, have faded, and others have taken their place. However, Midgley's resistance to the idea that a more masculine world is a better world is enduring and present throughout her work. This is one of the key takeaways from her critique of Western philosophy. We should resist the idea that hypermasculinism—in all its guises—is an aspirational ideal, in philosophy or life. There are more nuanced and integrated ways of living. And, furthermore, there are more nuanced and integrated ways of doing philosophy. Midgley long modeled this for us, ensuring that women's voices were heard in philosophy.

7

Science in Context

It's the exclusiveness that's the trouble with scientism.
And it's often an attack on the humanities.[1]

Science, which involves the systematic study of the natural world, plays a major role in how we envision the world and conceive of reality. At the same time, science is shaped and directed by the larger mythic structure of our lives. Thus, it requires an understanding of the wider map or context of human life. Midgley argues that, if we attend to science's functions, limitations, and connection to our wider world-pictures (myths or imaginative visions), then we can use it wisely and sensibly. If we do not, then we run the risk of misunderstanding—and, worse, misusing and distorting—science. Certain accounts of science, including scientism and reductionism, can do just this. Scientism can excessively isolate and glorify science while reductionism can promote overly simplistic, one-sided explanations of a complex world. These accounts of science typically neglect other important variables, both within and outside of science, that can inform and guide a more nuanced appreciation of science in context: in the case of scientism, the connection of science to our wider purposes and forms of knowledge, and, in reductionism, the connection of one-sided explanations with other complementary accounts. Of course, explanations of "reality" require an accounting of the powers and limitations of any type of human inquiry. This does not mean—as some skeptical opponents to scientistic and reductionist ideologies have suggested—that science needs to be cut down to size as a mere social construction. Rather, it means balancing and taking seriously the insights of both realism and skepticism in conceptualizing science as an essential element in understanding the world.

Conceptualizing and contextualizing science

Conceptions of science typically involve an emphasis on some combination of its unique systematic approach (methodology) and its body of knowledge (epistemology). Of course, internal understandings of the nature and function of science vary, largely because science is not an undifferentiated monolith.[2] Midgley reminds us that it includes a wide range of approaches, methods, functions, and assumptions that vary greatly between the many disciplines that we know as "scientific."[3] Experimentation is more common in chemistry than in cosmology, and observation is more fundamental in ethology than in mathematical physics. Studying people in psychology and sociology is different from studying subatomic particles in quantum mechanics. Inductive approaches and assumptions factor more in microbiology than in the more deductive theoretical physics, though all sciences use both induction and deduction.[4] In other words, the sciences share common ground while exhibiting rich variety *appropriate to* the given discipline and its subject.

Our knowledge of the world is heavily dependent on science, but there are many kinds of knowledge, each of which contributes to our understanding of ourselves and the universe. In this way, the scientific disciplines and their unique constitutions, along with the complementary accounts of well-developed "ways of knowing" in history, literature, art, and so on, contribute to our collective knowledge.[5] None of these approaches is infallible nor the standard by which to judge all others, as each has methods and explanations appropriate to its subject and aims. All patterns of knowledge, Midgley writes, function more like a complex network or "a city of different interconnected buildings, than . . . a single enormous building piled on a single foundation-stone."[6]

Midgley employs numerous such images to draw out the connection of scientific knowledge to other forms of knowledge. Much like a large, dimly lit aquarium, we can look at the world from many distinctive viewing windows.[7] These windows include the various scientific disciplines—prominently featuring physics, biology, and so on—but also those outside of the sciences, including history, literature, and philosophy. We can track the various inhabitants in the aquarium, though we must regularly "dash about" from window to window if we want to connect the data in relevant ways. We may never perfectly understand the aquarium and its contents, but we can move toward the fullest picture possible by employing all our available standpoints. No one approach has a walkover—understanding animals, for instance, requires biologists, ethologists, philosophers, animal trainers, veterinarians, poets, and so on.

Midgley argues that philosophy is significant in this connective process and is particularly helpful in dealing with large-scale conceptual issues and making connections between the many aspects of our thought. While some may wish to cordon philosophy off from scientific thought—Lawrence Krauss, for instance, remarks that "the worst part of philosophy is the philosophy of science. . . . It has no impact on physics what so ever"—this separation is untenable and undesirable.[8] This is because conceptual thinking and numerous philosophical assumptions are involved in the practice and interpretation of science, and there is risk involved if they are unrecognized. Science developed—and continues to develop—on sets of philosophical concepts and assumptions about the regularity of nature, the existence of an outside world, the reliability of our senses, the value of induction and empiricism, the principle of parsimony, and so on. These concepts and assumptions connect together in the larger structure of scientific thought and must be monitored, scrutinized, and refined—by both internal and external critics—because of their impacts. Midgley writes:

All science includes philosophic assumptions that can be questioned and those assumptions don't stop being influential just because they have been forgotten. They lie under the floorboards of all intellectual schemes. Like the plumbing, they are really quite complicated, they often conflict, and they can only be ignored so long as we don't happen to notice those conflicts. When the conflicts get so bad that we do notice them, we need to call in a philosophic plumber—not necessarily a paid philosopher, but someone who knows how the philosophical angle matters.[9]

By posing and exploring questions related, for instance, to assumptions about methodological, epistemological, and metaphysical naturalism—that is, how we should *study* nature, what we can *know* about nature through this study, and what philosophical claims we can make about the *nature* of nature—scientists and philosophers can engage in productive dialogue about how to best conceive of the world and our place in it. Concepts such as time, space, matter, causality, and life, furthermore, can require as much philosophical as scientific thinking.[10] Einstein, Heisenberg, and Darwin are just a few of the scientists who have engaged in this dialogical and integrative approach, forging new ways of understanding their scientific findings in the whole of life. Midgley argues that this pursuit, which overlaps the boundaries of science and metaphysics, can be done in better and worse ways, but it is ultimately unavoidable.[11] This is what those such as Stephen Hawking and Leonard Mlodinow miss in making metaphysical claims about reality while pronouncing that "philosophy is dead."[12]

Scientism and reductionism

With these considerations in mind, Midgley urges us to be pro-science and anti-scientism.[13] Scientism (or *scientisms*, since there are many variations) is not science, but rather an ideology that tends to isolate and exalt the "*idea* of science on its own."[14] A variation of this is demonstrated by the chemist Peter Atkins:

Although poets may aspire to understanding, their talents are more akin to entertaining self-deception. They may be able to emphasize delights in the world, but they are deluded if they and their admirers believe that their identification of the delights and their use of poignant language are enough for comprehension. Philosophers too, I am afraid, have contributed to the understanding of the universe little more than poets. . . . They have not contributed much that is novel until after novelty has been discovered by scientists. . . . While poetry titillates and theology obfuscates, science liberates.[15]

This isolation—which, it should be noted, need not involve "any real understanding of [science's] methods"[16]—can cut science off from "the rest of thought . . . treating it as a victor who has put all the rest out of business."[17]

Midgley argues that this move is unrealistic. It is more than reasonable to properly value and appreciate science, as "we can never feel too much reverence for science, or for any branch of knowledge."[18] This is because "knowledge is indeed wonderful and should be revered."[19] Yet, we can and must "value and celebrate scientific knowledge without being dragooned into accepting propaganda which suggests it is the only thing that matters."[20] "Human thought," she writes, "operates as a whole. It is an ecosphere, a vast and complex landscape."[21] Science, in this picture, is, again, not an isolated monolith or compartment but "a large, thickly wooded area comprising many sciences, an area that merges into those around it."[22] Various scientific disciplines draw on each other, as well as from concepts in the wider landscape of thought, including from areas outside of science. Biology, for instance, includes philosophical elements related to the concept of life and historical elements relating to the change and evolution of life.[23] In this way, science not only influences but also grows out of "the pre-existing social soil and is wholly dependent on it."[24] Scientism (or at least the kind of scientism that Midgley is most concerned with) largely ignores this landscape, isolating and elevating science as a self-contained standpoint and, moreover, the only rational way to see the world. All the while, it tends to perpetuate imagery such as "science versus literature, intellect versus imagination, analysis versus synthesis, expert versus amateur, man versus woman."[25]

Scientistic ideologies, furthermore, can unite with one of Midgley's other targets: reductionism. These one-sided accounts emphasize a favored set of explanations—typically materialist ones within the physical sciences—while downplaying other explanations and accounts of knowledge. Reductionism is seductive because it holds the promise of simplifying immense amounts of data and descriptions of the world into a clean, tidy order. This appeals to our general desire for order as well as the idea that the true essence of things can be found in their simplest underlying elements. However, this pursuit, and the certainty of its convictions, can become exaggerated. Reductionism is one way among others of seeing the practice of science. Some ways of doing science require a healthy dose of reduction (the physical sciences) and others less (the life sciences). In other words, understanding some things, like atoms and molecules, requires reduction. The problem with reductionist ideology, in the forms Midgley critiques, is that its proponents assume or insist that reduction is essential to *all* explanation: of thoughts, minds, people, and so on. Midgley's favored example here is Francis Crick, who claimed: "You, your joys and sorrows, your memories and your ambitions, your sense of personal identity and free-will, are in fact *no more than* the behavior of a vast assembly of nerve-cells and their attendant molecules."[26] Reductionist claims such as this represent metaphysical decisions as much as anything else.[27] This is a matter of selection and emphasis, combined with a healthy dose of pleasure in fulfilling the role of a debunker or unmasker.

Things of increasing complexity, such as organisms or consciousness or ecosystems, call for increasingly complex descriptions. Midgley argues that a pluralism of maps or lenses onto a given topic provides this rich set of complementary, rather than warring, descriptions of reality. This convergence of approaches can happen between scientific disciplines and can also involve disciplines and thinking outside of the sciences. Just as a chemist, a botanist, a historian, and a poet would provide differing but complementary descriptions of a tree, different forms of knowledge do different jobs, both within and outside of science.[28] A tree is more than its chemical makeup, just as a human being is more than "only £5 worth of chemicals."[29] One single explanation does not have to be fundamental or final.

For Midgley, "discriminating or intelligent pluralism," as opposed to reductionism—or, at least, reductionism *alone*—allows for the multiplicity of sources required for understanding phenomena in the world in appropriate ways.[30] Science, in other words, must not be too one-sided in fitting the best systems of thought to the best questions. Ideally, approaches converge, rather than compete, in piecing together the fullest knowledge. And, she notes, this *is* what good science does. Though the drive to a unified field of inquiry is understandable, as it reflects our general desire for order and simplicity, the tidiness sought in reductionist accounts—tidiness at any cost—can be misguided, partial, and ultimately distortive of our vision of the whole.

One of the most consistent metaphors that Midgley uses in getting across this vision of the whole is cartographic.[31] Atlases, for instance, provide a number of maps, such as physiography, vegetation, climate, and geopolitics.[32] Though we are more likely, these days, to pull various maps or overlays up on a computer or phone than in an atlas, this imagery still pertains. Each map or perspective provides a different framework which complements, rather than contradicts, the others. If we are going to understand Australia or Greenland, we need all the maps available—they are all partial and incomplete, and, in turn, work best as a whole. Likewise, mapping ecosystems involves relating differing ways of seeing organisms, inorganic entities and collectives, weather patterns, and so on. Reduction to one explanation or one map leaves out a variety of elements that are ultimately irreducible to others. These explanations and the language used to describe them—which is often metaphorical—must be balanced, Midgley argues, much like how Darwin alternatively drew on the metaphors of the profusion and wastefulness of nature in discussing natural selection.[33]

Myths and meaning

Metaphorical language is more than fitting for conceptualizing the world since science itself is more than just an important tool for understanding the world. Science grows out of, and connects with, some of our most central aims and purposes in life. In this, it shares interesting features and borderlines with other aspects of life, such as religion (a sense of awe and wonder) and poetry (expression of genuine interest in things other than ourselves).[34] Science, in this connective sense, is not an inert or detached institution. It generates world-pictures (myths or imaginative visions) and meaning that dramatically shape our everyday lives by influencing how we think of ourselves and the world around us.[35]

Myths, for Midgley, are generalized representations of the world. They are not falsehoods or lies, in the colloquial sense, but "imaginative networks of powerful symbols that suggest particular ways of interpreting the world."[36] They are, to some extent, both descriptive and prescriptive, reflecting the world while also influencing how we act in it. As such, science informs myths while also, to some extent, being influenced by them. Recognizing these underlying visions is therefore crucial, as they can be devised and disseminated in better and worse ways. Masculinist, competitive, and mechanistic myths of the natural world, for instance, draw on a whole range of interconnected ideas and values and have real impacts on science and its interpretation. Midgley warns that if science is conceptualized as an isolated phenomenon, quarantined away from our generalized visions and ideals of the world, it is prone to let those very visions sneak in and undermine the integrity of science. This is especially the case with myths that masquerade as science, such as the myth of progress, the myth of meritocracy (that the best scientists "reach the top"), or the meta-myth that the Enlightenment eliminated myths once and for all in favor of purely scientific language and descriptions.[37]

Myths engender the dramas—or powerful stories—through which we depict and live out our lives. This matters because facts coalesce into dramas that can, potentially, distort science:

Facts will never appear to us as brute and meaningless; they will always organize themselves into some sort of story, some drama. These dramas can indeed be dangerous. They can distort our theories, and they have distorted the theory of evolution perhaps more than any other. The only way in which we can control this kind of distortion is, I believe, to bring the dramas themselves out into the open, to give them our full attention, understand them better and see what part, if any, each of them ought to play both in theory and in life. It is no use merely to swipe at them from time to time, like troublesome insects, while officially attending only to the theoretical questions. This will not make them go away, because they are a serious feature of life.[38]

This dramatic influence, as such, selectively lights up aspects of scientific visions. Individualist and competitive dramas, for instance, tend to spotlight instances of violence as vivid representations of the larger drama of natural selection. The problem, for Midgley, is when these dramas are one-sidedly sensationalized and essentially take over.[39] Though there are certainly egoistic, competitive, and violent elements in the natural world, there are also elements of toleration, cooperation, symbiosis, and peaceful interaction. In other words, the dramas must be balanced in our visions of the whole.

If these dramas and their respective myths are not balanced, tempered, or otherwise harnessed into productive and cautious scientific processes, they risk undermining science itself. Scientistic dramas are particularly prone to reducing and over-dramatizing topics such as evolution (see "Religion and Science"). Midgley thus advises that the "myth-garden" be regularly weeded.[40] One particularly weedy myth that Midgley critiques throughout her work is an unrealistically atomistic and mechanistic vision of the world. She fiercely opposes these visions as threats to how we see the wholeness of the self (see "Human Nature"). In its most basic form, mechanistic atomism is the idea that "the only way to understand anything is to break it into its ultimate smallest parts and to conceive these as making up something comparable to a machine."[41] Dating back to the Epicureans, atomism has helped to explain many aspects of the world in innovative ways. Everything from bacteria to medicine to economies has been seen as describable through atomistic methods—that is, through analysis of the interactions and aggregations of discrete units in the world. But Midgley argues that this is "one possible interpretative pattern among many," useful in some circumstances and not in others.[42] She, again, is critical of atomistic and individualistic myths because they focus on the parts but often ignore their relations to the whole. In other words, we have to put things in the perspective of the bigger picture. Organisms, ecosystems, economies, and other complex systems require more than analysis of their component parts in isolation. They require holistic approaches that understand the parts in context and how they relate to each other and the whole. The nature and behavior of bacterial biofilms, for instance, are different than those of individual bacterium.[43] If we want to cure diseases caused by these biofilms, we probably need to understand the individual and collective levels of bacterium involved. "*Holism and atomism*," in this way, "are not warring alternatives. They are complementary aspects of all scientific enquiry."[44]

Our world-pictures (a term Midgley uses interchangeably with myths) represent the underlying systems of thought through which we view the world.[45] These shared world-pictures have a powerful influence on the way we live, and their influence extends into science by filtering facts into a meaning-giving structure.[46] Facts are rarely, if ever, brute or raw data but imbued with values and concepts. Accordingly, we are never wholly neutral in selecting, interpreting, and classifying data, all of which fall into preexisting dramas and world-pictures.[47] Meaning arises, in part, in how these facts are integrated and ordered within our world-pictures.[48] This does not necessarily mean that we neatly and tidily construct our ways of seeing the world into a comprehensive pattern. Rather, it points toward a value-laden network through which facts are understood in relation to the "central region" of meaning in our lives. The closer facts and our knowledge of the world connect with this central region, the more significance and meaning they have. World-pictures, as such, are the connective systems through which we order our thought. They are the background for all of our pursuits, including science. For this reason, the whole structure on which we base our knowledge must be thoroughly examined and recognized in the context of our wider purposes and motives.

World-pictures are also relatively flexible and responsive to a range of factors, such as self-reflection and social influences. They are prone to changes and gradations over time, and meaning can occasionally be radically reoriented; for example, when facts about the origins of the universe or life on earth don't fit with existing world-pictures (see "Religion and Science"). In our more reflective moments, we seek "greater completeness" and coherence by sorting through and reconciling elements that clash.[49] However, this dynamic is not always smoothly responsive and reflexive. Since world-pictures are guided by our imaginations, motives, and purposes in life, they are capable of eclipsing all other world-pictures. Vibrant dramas can take over and blind one to faults within their visions. This calls for "responsible objectivity," or "the difficult task of becoming aware of one's world-picture, doing all one can to correct its more obvious faults."[50] This is especially important for scientists, who might assume that their world-pictures are purely detached and neutral lenses onto the world. Yet, she writes, "like those who argue usefully on any other subject, they do their best work not by being neutral but by having strong preferences, being aware of them, criticizing them carefully, expressing them plainly and then leaving their readers to decide how far to share them."[51] Scientists have great power and responsibility not only in their scientific attitude and selection of topics but also in the imaginative visions they express, endorse, or otherwise support, whether intentionally or not.[52]

Subjective and social influences

Midgley's emphasis on personal factors in scientific processes should come as no surprise. After all, scientists are people and science is a human practice. This premise is relatively uncontroversial. Yet, some may regard science as somehow detached from any human pursuits and immune to subjective influences. Subjectivity may have a role in the arts or the humanities but not in the sciences. This assumption makes sense because science developed as a way to *minimize* the biases and whims of people and institutions in the systematic pursuit of knowledge. Our understanding of chemical reactions or tapeworms or climate change should not be determined by wish fulfillment or the interests of political or religious groups. This said, science, again, is not detached from all human aims and influences. Otherwise, why would we do it?

Individual and social influences can factor into the direction and, to an extent, findings of science. The world-pictures of scientists, which are laden with values and presuppositions from outside of science, play a vital role in what they select and emphasize: what they consider important.[53] This is not necessarily a bad thing, as it demonstrates how science is inherently responsive to real human needs. In this soft sense, bias and partiality are simply unavoidable. Yet, one can recognize and effectively balance these subjective factors in useful ways, and scientists do this all the time. Refusal to acknowledge subjectivity, on the other hand, can be paradoxical, as one simply overlooks or confuses one's own personal influences and motivations as internal to the practice of science. One might, for instance, confuse one's personal ideology or values—climate change skepticism, creation science, fatalism—with objective science.[54] Even Einstein's personal views—relating to assumptions of God and the objective order of the universe—famously led him to reject the indeterminacy of quantum mechanics: "He does not play dice with the universe."[55]

Objectivity, in this way, is an *ideal* rather than a given. This is not a bad thing and does not tarnish scientific pursuits. Ideals are necessary for any venture, as we constantly recalibrate and refine our practices in pursuit of lofty goals. For Midgley, it is thus perfectly reasonable to pursue objectivity while at the same time being reflective about one's subjectivity.[56] Objectivity, in this sense, might look like "keeping irrelevant biases out of science. It means being fair to one's opponents, not letting one's political views interfere with one's reasoning, and, again, not letting one's pride stop one from giving proper attention to earwigs or tapeworms."[57] Well-trained and reflective scientists model this practice. They are skilled in the critical "background thinking" that positions science and the scientist within a wider context.[58]

Toward a balanced realism

Social factors also influence science. In the words of the neuroscientist Santiago Ramón y Cajal: "Like all mental activities, the accomplishments of the scientist are heavily influenced by the physical and moral environments around him."[59] Ideas, naturally, do not spring up in isolation. Rather, they grow up organically out of specific social needs, concerns, and contexts. This, like most human practices, can be for good or ill.[60] As such, some skeptics have argued that science, as a social construct, might be nothing more than a reflection of the wishes and desires of the scientific community or society at large. Science, in this picture, might be more of a mirror of ourselves than a window onto the world. In other words, science just reflects and reinforces our world-pictures.[61]

Skeptical perspectives such as these are typically reactions to their antithesis: simple or naïve realism. This is the notion we can, as it were, look directly on the world as it *really* is, without the filter of the many influences that shape our experience of the world: our mental faculties, world-pictures, personal and social factors, and so on. Midgley, again, rejects both the skeptical and realist positions in their extremes. Each contains its own partial truths. Skeptical positions draw attention to the nature of science as a practice embedded in human life and culture. Realists, on the other hand, draw attention to fact that we do have some ability to track and describe the world as it really is. The problem for each, however, is that they rely too heavily on the assumption that there can be such a thing as *final* or *perfect* knowledge (see "Morality"). Naïve realists assume that this knowledge can and should be attained and thus discount or ignore the important insights of skeptics. Skeptics, on the other hand, assume that such knowledge cannot be attained, and thus dismiss its pursuit. Midgley insists, however, that there is nothing in human life or science that requires this finality of knowledge.[62] She argues that a better conclusion in the realism-skepticism dialectic is that science may have some limits, but, rightly conceived, it does provide an invaluable set of windows into the aquarium, so to speak. The aquarium may be dimly lit, but we do have the ability to peer in and gather some reliable observations about its interior. This is regularly confirmed not just in the explanatory power that science provides but also in its ability to make testable predictions and practical applications. Even an incomplete causal explanation of the human immune system, for instance, allows for countless lives to be saved through the practice of modern medicine.

Midgley thus argues that skepticism and realism should not be seen as warring alternatives, but rather as two poles between which our understanding of the world must be respectably balanced.[63] Ultimately, her position shades much closer to the realist than the skeptic pole (there *is* an aquarium/world out there, however dimly we may view it):

> Of course our own individual point of view makes a great difference to how we see things. But that difference is much better described as selection than as construction. It is quite true that, when we look at the Himalayas, every one of us sees them differently. But none of us can think them away, nor put them there in the first

place. All perception takes in only a fraction of what is given to it, and all thought narrows that fraction still further in trying to make sense of it. This means that what we see is real enough, but it is always partial.[64]

Science in context

Midgley suggests that we should consider any inquiry and its knowledge on a sliding scale of better and worse. The idea of perfect methods and perfect knowledge is probably unrealistic and can distract from working hard at and refining what we *can* do, which is being self-reflective, cautious, curious, and fastidious in working toward the most complex maps we can devise of the world. Skepticism has a proper role within this process, since scientific inquiry is, after all, inherently skeptical: question, test, review, and revise. Knowledge, here, is always provisional, though some things are more certain and "final" than others.

In this way, the methods, process, and spirit of science push back against scientistic and reductionist accounts, which not only isolate science from its wider context and purposes but also reduce explanations of the world to a chosen set of fundamental and often *final* descriptions. This largely animates Midgley's efforts to contextualize science. Yet, there is more. These near-prophetic exaltations and distortions of science can further the narrative of the warfare between science and other areas of life—religion, in particular—while paradoxically fulfilling some of their functions. In these cases, scientism (and related accounts) can even *look* religious. I cover this, along with Midgley's general examination of the relationship between religion and science, in the following chapter.

8

Religion, Science, and Complexity

I suppose in a very admirable and harmless way, I take science to have a religious aspect. That is, it involves people responding to the vastness of the universe.[1]

The relationship between religion and science is complicated, and distortive accounts of each tend to belie their complexity. This relationship is especially obscured when certain forms of scientism—in their isolation and glorification of science—*look* religious. Of course, this all depends on what religion looks like, which is complicated in itself. Yet, with the intricacies of both science and religion in mind, Midgley embarks to disentangle some of the conceptual issues involved in their relationship, which entails working through some of their similarities and differences. This investigation does not amount to a project of merely tracing parallels. Instead, it is largely an attempt to maintain the integrity of *both* science and religion, lest distortions of each obfuscate their functions. Midgley ultimately seeks to contextualize and, to an extent, reconcile the more damaging ways in which religion and science have been portrayed and related. In the process, she draws attention to their dual contributions to our visions of the world, and how these can go awry.

Conceptualizing and contextualizing religion

The complexity of religion is reflected in the diverse range of methods and approaches through which it is studied.[2] Historically, the study of religion has been predominantly belief-centric, focusing on the more creedal and intellectual sides of religion. However, contemporary religion scholars are quick to point out that this approach to religion can overlook the rich tapestry of religion "on the ground," where people live out and experience religion. For many, this lived aspect is as or more significant than its intellectual or conceptual elements.[3] "Religion" itself is an elastic term, partly due to its nature as a recent, academic, and largely Western way of compartmentalizing many diverse elements of life into one tidy domain.[4] Conceptualizations of religion tend to fall between the poles of essentialist (what religion *is*) and functionalist (what religion *does*) accounts. Most contemporary approaches highlight some combination of these aspects in the effort to not reduce religion to any one thing. More recently, due to the variability of the term, some theorists have employed a Wittgensteinian "family resemblance" approach to religion.[5] Here, religion has no fixed essence or function but exhibits widely diffused and overlapping patterns of each across a range of traditions. Of course, some of these patterns and resemblances are shared by nonreligious or secular traditions—for instance, sets of rituals and ethical norms are present in various political ideologies and even in sports fandom. Though some consider this a limitation of the family resemblance approach to religions, others see it as an advantage because it is, in the end, more descriptive of the complex and fluid boundaries of religion and religious-looking phenomena in the world and allows us a wider context for understanding its various patterns across many forms of life.

Midgley's approach to religion looks like this family resemblance approach, though she also displays elements of functionalism. She highlights the plurality of religious features, both within religions and in society more broadly, and how they function in human life. Religions contain elements that are often connected with morality, the arts, politics, and science, for instance, and its component parts often shift and rearrange over time. This is especially the case when historical and cultural forces decenter the relative position of religion within a society—much like the church in Western Europe—which can lead to interesting and intriguing patterns and combinations of religious elements over many forms of life. Religions, in this way, "develop out of a number of separate components—human motives and tendencies which can grow together and be combined in various ways."[6]

Myths

In mapping the terrain of religion, Midgley focuses on one element in particular—myths, or world-pictures—because of their centrality in virtually all aspects of human life. Myths, again, "are imaginative networks of powerful symbols that suggest particular ways of interpreting the world" (see "Science").[7] She reiterates that myths are not necessarily true or false, in the colloquial sense, but deal with truths about the world through imagery and symbolic visions. Crucially, they play a central role in shaping meaning, which involves the way in which facts coalesce around a central core of "significance or importance."[8] Religions and religiosity often seem to grow out of the awe, reverence, and faith that these myths express. Though this is not religion per se, these elements are usually present in some form or fashion in religious ways of life.

All of this matters, whether we consider ourselves religious or not. Whatever our lives are like on the ground, Midgley reminds us that we all think. And, we tend to think in these mythic and metaphorical ways, which deeply shape how we see ourselves and the world around us. Ideally, our myths are responsive to the world around us, taking in a range of perspectives from various forms of knowledge (see "Science"). However, she notes that we often do not recognize the myths that underlie how we see the world. We might think we see reality clearly and directly—the world *as it is*. Though myths are central to our lives, we can be unaware of their shaping influence.

This isolationism and uncritical acceptance of certain myths and their attendant dramas—typically reductive ones—is a central problem in distortive accounts of science. Myths such as the "omnicompetence of science" promote a vision of the world in which science can explain everything and provide for all our needs, obviating the need for other forms of knowledge and even ways of living (such as religion). For this reason, Midgley argues it is important that we recognize the myths themselves, as well as the roles they play in shaping our lives. If we are reflexive, we can collectively interrogate and rebalance unrealistic and potentially harmful myths and dramas.[9] If not, they threaten to distort not only science and religion but our lives as a whole.

The warfare of religion and science?

The relationship between religion and science is regularly oversimplified and caricatured as one of conflict. This image is promulgated and sensationalized in the media—"God vs. Science"[10]— and has a strong grip on the public understanding of this dynamic. Though there are certainly conflicts between different types of religions and science, contemporary scholarship on this relationship points toward numerous modes of interaction, including harmony, independence, dialogue, integration, and so on.[11] These "models" of the relationship between religion and science aim, in varying measures, at description (how they *do* interact) and prescription (how they *should* interact). Midgley does not expressly support or construct any one particular model of religion and science in her critique of scientism. This is not really her goal. Rather, her position points more toward a generalized examination and understanding of the complexity and interconnection of religion and science in context.

Midgley is primarily opposed to the notion that religion and science are locked in an epic war in which a victor will eventually emerge once and for all. This warfare or conflict model emphasizes prominent historical examples—such as Darwin versus the religious orthodoxy of Victorian England and Galileo versus the Catholic Church—to illustrate the inevitable cosmic clashes that arise between religion and science. This seductive narrative, however, isolates and exaggerates instances of conflict over other modes of interaction. There are usually more dynamics involved in these historical episodes than one simple account can provide. Conflict and division may occur not only *between* religion and science but also *within* religion and *within* science. Some resistance to Darwin came from within the scientific community (Richard Owen) and some support for Darwin was from the clergy and other prominent religious leaders (Charles Kingsley). Both scientifically and religiously minded people (who were often one and the same) grappled with the implications of Darwin's theory. Midgley argues that this was not necessarily because natural selection threatened faith per se (though it did for some) but, rather, because it was perceived as a threat to the dignity of human beings (see "Human Nature").[12]

Historians have examined the rich historical landscape of such conflicts while muddying the waters of the conflict metanarrative and the idea that there is something about science and religion in themselves that necessitates this inevitable battle. Contemporary historians of religion and science now argue that complexity is probably a more representative description of this relationship.[13] The warfare thesis extrapolates too broadly from *aspects* of isolated cases to the whole of religion and science. Clashes between science and specific sects or denominations of the Christian tradition, in particular, are taken to be representative of the *whole* of the relationship. Given the variety and differences within traditions—note the wide disparities within Protestantism alone—these cases cannot even be seen to characterize the relationship between science and the Christian tradition, not to mention the diversity of religions around the world.[14] Tibetan Buddhism and Sufism do not necessarily follow the same patterns as Evangelical Christianity or Daoism. Numerous factors—personal, moral, philosophical, historical, economic, political, and so on—are typically involved in the interactions of *religions* and science. Or, perhaps even better, between our individual world-pictures and scientific facts. Stephen Jay Gould, for instance, highlights how both Darwin and T. H. Huxley, who became potent symbols of the conflict thesis, opposed certain religious ideas for personal and moral reasons, not necessarily for scientific ones, though each realm contributed to their overall outlook.[15] Others have indicated that Galileo, another prominent symbol of the warfare thesis, was perhaps more of a victim of the Catholic Church's movement to restore its authoritative role in biblical interpretation in the wake of the Protestant Reformation than a casualty in the religion and science war: it was the Church's and not Galileo's role to dictate how geocentric scriptures were to be interpreted.[16] In other words, there is typically more complexity involved between "religion" and "science" than certain accounts let on, at both the institutional and individual levels.

Religion as science?

Yet, this does not mean that there are not real conflicts. Disagreements about how to read and interpret religious texts feature prominently in the warfare model of religion and science. This is most visible in traditions that emphasize the inerrant authority of sacred scriptures. Though different Christian communities have traditionally read the Bible in numerous ways based on form and content—metaphorically, allegorically, historically—more recent fundamentalist and literalist interpretations leave much less room for metaphorical readings. This form of interpretation underlies the predominantly American phenomenon of creation science (or scientific creationism).[17] Advocates of this position pit the Bible, especially the Genesis narrative of creation, against mainstream science, claiming that the biblical record of creation is tantamount to a literal scientific description of the origins of the cosmos and life on earth. This type of approach to biblical interpretation was most infamously spotlighted in the "Scopes Monkey Trial," where creationists clashed with "evolutionists" over the teaching of the theory of evolution in public schools in the United States. This attempt to scientifically explain the world through religious texts is where notions of the conflict of science and religion are most accurate for Midgley. The myths, in such cases, take over and become unbalanced. In fact, the myths are often not recognized *as* myths (which, again, does not mean that they are false). If scientific facts conflict with these myths or world-pictures, they are, in turn, either manipulated to fit the system or rejected entirely.

Thus, there are real examples of conflict, though Midgley argues that these conflicts should not be seen as templates for the whole of religion and science. This particular example of conflict occurs with a particular version of religion. Here, religion takes on some of the functions of science by attempting to explain the scientific facts in literal and historical terms (as opposed to the symbolic terms many Christian communities have employed).[18] This is precisely the type of interpretation that was opposed by Galileo, who claimed that the Bible is a spiritual rather than a scientific document.[19] He argued that if biblical facts seem to contradict scientific facts, then one or the other sets of facts is likely being misinterpreted, since truth cannot contradict truth. Galileo preferred to reinterpret these biblical facts in more metaphorical terms (as accommodated by God to human language), whereas creationists have favored more literalist accounts of the facts.

Complete separation?

Midgley argues that religion, in such cases, "has left its own sphere . . . to intrude on a scientific one."[20] She anticipates and largely aligns with the position of Stephen J. Gould, who insists that religion and science, ideally—that is, prescriptively—should not and do not clash. This is because religion and science occupy different domains or "magesteria" (areas of teaching authority).[21] The authority of science is in facts and theories related to the natural world, while religion (along with philosophy) has authority in the domain that deals with meaning, morality, and purpose. These separate domains, in their various ways, come together for the "fullest wisdom." Clashes only result when one domain tries to take on the functions of the other.

This approach, which Gould terms "non-overlapping magesteria" (NOMA), protects the domains of religion and science from manipulation and distortion. This is one of Midgley's central aims in examining such distortive accounts of both religion and science. Yet, this compartmentalization into separate spheres may not be as tidy and clean as it first appears, especially in the borderline areas. These airtight spheres may be too rigid and belie some of the shared features of religion, science, philosophy, aesthetics, and other aspects of life which do not actually exist in complete isolation from each other. Rather, they share borderlines with and, in some places, can even overlap each other. This is primarily because all of these areas, though largely distinct, still grow out of human life and culture. Underneath the divisions between these different areas of life is a common source in our shared aims and purposes, our motivational structure. These areas are all tied together through our world-pictures and contribute to life as a whole. Midgley highlights that the distinctions, in the end, between religion and science are "not as plain, nor the relation between them as simple as is often supposed."[22] This understanding does not mean that religion and science—nor any other of these aspects of life—are the same. Far from it. Rather, the point is that their shared elements are more significant than the compartmentalization approach suggests. She then conceives of religion and science not as inherently clashing, nor as completely isolated from each other, but as complementary aspects of life as a whole that have a common source in human life and motivation. Our world-pictures are so complex because they draw on all of these sources while likewise exerting their own influence on our lives.

Science as religion?

Understanding the overlapping elements of religion, science, and other areas of life may help us understand why these complementary areas may, at times, eclipse or take over the others. Science provides powerfully symbolic insights and visions on our origins, nature, and destiny (see "Science"). It is also motivated by many of the same elements that spur religiosity—awe, reverence, and curiosity—and involves "people responding to the vastness of the universe."[23] Problems can arise, however, when science is isolated, emphasized, and venerated to the exclusion of other aspects of life. Midgley argues that this is roughly what happens with scientism. (This, it should be noted, is not just an issue when it comes to religion, but for any form of life and knowledge.) Certain accounts frame the scientific domain as the worthiest pursuit in life, capable of giving meaning, value, and even insight into the nature of questions often considered religious. Much like creation science, this approach attempts to subsume or conquer the other domain—this time religion—and imagines its myths as *final*. In the process, scientism, like creation science, distorts both domains.[24]

This distortion is especially apparent with unbalanced myths stemming from scientific theories. The myths surrounding evolution are particularly enthralling. "The theory of evolution," Midgley writes, "is not just an inert piece of theoretical science. It is, and cannot help being, also a powerful folk-tale about human origins."[25] As "the creation myth of our age," it provides much of what religions traditionally offer: an origin story.[26] It shapes our thoughts, feelings, and actions by its symbolic power, much as religious myths have. This does not mean that evolution *is a religion*, but simply that it engenders potent imagery and myths that can be portrayed in ways that look religious.

Two episodic distortions of evolution looked particularly religious for Midgley. Contrasting dramas of teleological progress and competitive egoism represented "hypertrophied forms of cosmic optimism and cosmic pessimism," respectively.[27] The optimistic distortion of evolution was the portrayal of the linear and upward process of evolution, which culminates in something of a superhuman "omega point."[28] This "escalator fallacy" not only posited a human- or an intellect-centered *telos* in evolution but also advocated that humans should even intervene in this process to speed the arrival of this superhuman Utopia. The pessimistic distortion of evolution—or competitive egoism, especially in the form of Social Darwinism— is a familiar foe of Midgley throughout this book. It trumpeted the "survival of the fittest"—coined by Herbert Spencer—as the ultimate law of nature. Nature is "red in tooth and claw," and self-preservation and survival are the true underlying forces in human life. As egoists driven by the instincts to survive and reproduce, any altruism or other-regarding behavior is explained away as illusory, serving only the underlying principles of self-preservation. Evolution, in this picture, is the grim and *fatalistic* source of our struggles.

These distortions are less scientific than they appear and driven, in part, by false assumptions and ideology. The escalator fallacy, for instance, misuses the notion of purpose or direction in natural selection, emphasizing the near "chosen" nature of human beings. Darwin's theory resists the up-down directional hierarchy of an escalator or ladder, instead arranging "species in a radiating bush rather than on a ladder, accounting for all kinds of development, and also for some cases of unchangingness and of 'regression,' equally as limited responses to particular environments."[29] Social Darwinism is also, ironically, non-Darwinian. Though competition is, no doubt, pervasive in nature, it is but one element of the whole picture (see "Animals").

"Mutual dependence" and "harmless coexistence" are likewise widespread, and a social species such as ours must rely on more than just cutthroat individualism to develop and thrive (see "Human Nature").[30] Moreover, though Darwin did spotlight competition, he also emphasized altruism, cooperation, and the affective sentiments that stem from our deeply social nature (see "Morality").[31]

Midgley argues that these positions and their uses (which still linger) are clearly motivated by elements outside of science itself, typically in political, social, or even religious ideologies.[32] The myths guide the selective use of examples from the world in supporting the dramatic visions they entail, providing lively quasi-faiths in the process. One image among many is hyperbolized and exaggerated as *the* truth around which we should order our lives. The myths, whether in glorifying the future of humanity or competition, suggest a larger drama that gives us purpose or destiny, however utopian or dystopian. In this way, these distortions of evolution can parallel the religions that their prophets often wish to diminish or extinguish.

Midgley's discussion of religion and science demonstrates some of their parallels and overlapping boundaries. Yet, her comparisons are not simply about exploring similarities. While it is helpful to see the parallels between these spheres, she is more concerned with how the similarities can lead scientistic and other distortive accounts to look like, rather than simply parallel, religion. Her examination contains, to some extent, an effort to maintain the integrity of religion *and* science without completely isolating one or the other. Religion is the subject at hand, but one could probably extend this approach to science and philosophy, science and the arts, and so on.

Mixed antitheses

Midgley positions the warfare model of religion and science alongside the other antitheses throughout this book that threaten to undermine our total picture of the world. Religion and science are embedded within a larger system of mixed antitheses that tends to divide the world into warring alternatives. Features of the world have gradually become lumped together in a confused mix of oppositional—rather than complementary—dichotomies. These dichotomies are the effects of our efforts to organize the world in all its complexity into an intelligible order. This makes sense, as we want to manage the occasional chaos of life into workable categories. But, as the categories of religion and science demonstrate, classification and compartmentalization can indiscriminately create false oppositions, such as science *versus* religion, and fragment, rather than enrich, our understanding of the world.

As the post-Enlightenment vision of the world has become increasingly dichotomized and hierarchical, the general message underlying these oppositional antitheses is to "keep to the left" (with "the left" being the preferred side of the dichotomy: self versus others, human versus nature, male versus female, etc.) Some antitheses in this worldview are more justified than others. For example, science versus superstition, science versus partiality, science versus wish-fulfillment, and science versus dogmatism all make sense, as science must minimize personal and institutional influences and biases, insofar as possible.[33] In this way, "science versus" really is a reasonable way to view these troublesome elements in science.[34] However, other antitheses have become lumped together in less straightforward ways, representing a whole set of oppositions in which the contrasts are often clear, but the divisions are insufficient or problematic. These antitheses include such divisions as hard science versus soft science, empiricism versus rationalism, skepticism versus credulity, reason versus emotion, objective versus subjective, male versus female, specialism versus holism, and prose versus poetry.[35] At once it is apparent that, far from being oppositional or warring forces, these are "complementary elements in life and thought, both members of which are equally necessary, and indeed could scarcely be identified except in relation to each other as parts of a whole."[36] "We no longer want that truculent little 'v.' to divide them," Midgley writes, and efforts should be made to bring these elements together in more realistic ways.[37]

Envisioning the world

Whether we consider ourselves religious, spiritual, open-minded, skeptical, or whatever else, Midgley—herself an agnostic—argues that we all share in some aspects of religion. Being an agnostic

146 MARY MIDGLEY

or atheist or otherwise does not necessarily preclude a sense of connection to a greater whole outside oneself. This impulse, as vague as it may be, involves a longing for a "wider horizon" of life, whatever *it* may be.[38] The details of religious doctrines and dogmas tend to obscure this bigger picture for Midgley and so do distortions of religion *and* science. Coming to terms with the connective and outward-looking side of life entails grappling with issues of real significance or importance, since "a hunger for meaning is central to our lives."[39] Much like Darwin, Midgley understood the deep and lasting responses that come from the grandeur in the scientific view of the life. Here, the "real mysteries" of the cosmos can be grappled with in better and worse ways.[40] The visions that grow out of this process are natural responses to this confusing and messy world. However, visions that claim to have all the answers deceive us. The finality of unbalanced myths belies the complexity of the world, as well as the variety of imaginative tools that are at our disposal in conceptualizing it in its wholeness.

9

Afterword

"One World, But a Big One"

Mary Midgley passed away on October 10, 2018. Many of us, I think, had grown accustomed to thinking she might just live forever. We'd come to expect that she'd always be there when things got "dark and difficult."[1] Alas, this cannot be. Yet, Midgley's thought will live on. She had a profound ability to connect philosophy with life and its most important elements, wittily and incisively capturing it in all its complexity. And life, as she reminds us, "is essentially messy."[2] For this reason, she will always be relevant. To paraphrase Simon Blackburn: we will always need our Midgleys.

Midgley was able to see the manuscript of *Mary Midgley: An Introduction*, and we discussed it during our final visit in Newcastle. It was clear, in our conversations and correspondence, that she was still concerned with ways of seeing the world that divide complementary aspects of the whole into warring alternatives. This impulse to divide and fragment continually manifests in the way we see ourselves, others, and the wider world. The usual divisions—mind and body, reason and emotion, self and others, humans and nature—are all aspects, she reminds us, of the whole. This practice of breaking things up may have a function (for analysis of various types), but we must also bring things back together and connect them in the big picture. Midgley had a rare aptitude for this type of connective and holistic thinking, which underlies her critique and visions. She reminds us how big the world really is, and how deeply we are connected to it.[3]

These and other concerns are visible in Midgley's final work, *What Is Philosophy For?* The book, which was published the month before her death, is a fitting final statement on philosophy and its vital role in our lives.[4] She argues one last time that philosophy involves "the close relation between how we think and how we live."[5] This is all the more relevant as new challenges emerge in the world, challenges which often involve old problems in our patterns of thought. She spotlights scientistic impulses, including the near-religious reverence that is currently being directed toward artificial intelligence and the singularity. These and other problematic tendencies will continue to plague us if we do not recognize and correct them. It is in these circumstances, as she writes on the final page of her final work, that

> philosophical reasoning . . . will now become rather important. We shall need to think about *how* best to think about these new and difficult topics—how to imagine them, how to visualize them, how to fit them into a convincing world-picture. And if we don't do that for ourselves, it's hard to see who will be able to do it for us.[6]

Midgley is at her most prescient here, anticipating our troubles and reminding us that there *are* ways out of them. In these cases, we must look to ourselves, and to philosophy, to work out these tangles and find our way forward. Machines, she suggests, cannot do the job.

*   *   *

During our visits over the years, I was regularly struck by the way in which Midgley did philosophy. Brow-furrowed, deep in thought, she would light up and connect issues in the way described throughout *What Is Philosophy For?* After a cup of coffee and a biscuit or two, we would venture into the most interesting terrains, mapping the "conceptual geography" of the many topics covered in this book. I have tried to do justice to this process and this philosopher throughout, though any representation of her thought, as anyone who knew her would attest, will always be incomplete. In any case, the eight-year journey was well worth it. From here, I can only encourage the reader to delve deeply into her work themselves, where the "uniting spirit" will always be: "But look further!"[7]

Notes

Chapter 1: Philosophical Plumbing

1 Mary Midgley, Interview by Gregory S. McElwain, June 11, 2014, in Newcastle Upon Tyne, UK.

2 Mary Midgley, "Philosophical Plumbing," in *The Impulse to Philosophise: Royal Institute of Philosophy Supplement 33*, ed. A. Phillips Griffiths (Cambridge: Cambridge University Press, 1992), 139.

3 Midgley, Interview by Gregory S. McElwain, June 11, 2014.

4 Midgley, "Philosophical Plumbing," 139.

5 Ibid., 150.

6 Ibid., 140.

7 Mary Midgley, *Beast and Man: The Roots of Human Nature* (London: Routledge, [1979] 2002), xxxiv.

8 Ibid., 167.

9 Midgley, "Philosophical Plumbing," 139.

10 Ibid., 145.

11 Midgley, *Beast and Man*, 167.

12 Midgley, "Philosophical Plumbing," 141.

13 Mary Midgley, *Utopias, Dolphins and Computers: Problems of Philosophical Plumbing* (London: Routledge, 1996), 10.

14 Mary Midgley, *The Myths We Live By* (London: Routledge, [2004] 2011), 1.

15 Mary Midgley, *The Ethical Primate: Humans, Freedom and Morality* (London: Routledge, 1994), 19.

16 Ibid., 19.

17 Ibid., 18.

18 Mary Midgley, *Science as Salvation: A Modern Myth and Its Meaning* (London: Routledge, 1992), 14.

19 Mary Midgley, Interview by Gregory S. McElwain, May 29, 2013, in Newcastle Upon Tyne, UK.

20 Mary Midgley, *Wickedness: A Philosophical Essay* (London: Routledge, [1984] 2001), ix.

21 Midgley, *Beast and Man*, xiii.

22 Midgley, *Utopias, Dolphins and Computers*, 47.

23 Mary Midgley, *The Owl of Minerva: A Memoir* (London: Routledge, 2005), 131.

24 Midgley, *The Myths We Live By*, 223.

25 Midgley, *Utopias, Dolphins and Computers*, 44.

26 Midgley, *The Owl of Minerva*, 65.

27 Midgley, *Science and Poetry* (London: Routledge, [2001] 2006), 183.

28 Mary Midgley, Interview by Gregory S. McElwain, May 23, 2012, in Newcastle Upon Tyne, UK.

29 Ibid.

30 Jane Goodall, "Foreword," in *Science and the Self: Animals, Evolution, and Ethics: Essays in Honour of Mary Midgley*, eds. Ian James Kidd and Liz McKinnell (London: Routledge, 2015), xi–xii.

31 Midgley, Interview by Gregory S. McElwain, May 23, 2012.

32 See Ian James Kidd and Liz McKinnell, "Introduction," in *Science and the Self: Animals, Evolution, and Ethics: Essays in Honour of Mary Midgley*, eds. Ian James Kidd and Liz McKinnell (London: Routledge, 2015), 1–15; Benjamin Lipscomb, "'Slipping Out Over the Wall': Midgley, Anscombe, Foot and Murdoch," in *Science and the Self: Animals, Evolution, and Ethics: Essays in Honour of Mary Midgley*, eds. Ian James Kidd and Liz McKinnell (London: Routledge, 2015), 207–23; David Midgley, ed., *The Essential Mary Midgley* (London: Routledge, 2005); Midgley, *The Owl of Minerva*.

33 See In Parenthesis at http://www.womeninparenthesis.co.uk/, dedicated to the work of Anscombe, Midgley, Murdoch, and Foot. See forthcoming work *The Quartet* by Clare Mac Cumhaill and Rachael Wiseman (2022).

34 Midgley, *The Owl of Minerva*, 162.

35 Ibid., 192.

36 Midgley, *The Owl of Minerva*, 130.

37 Mary Midgley, *What Is Philosophy For?* (London: Bloomsbury, 2018).

38 A living bibliography of Midgley's work can be found here in conjunction with the Midgley Archives at the Durham University: http://reed.dur.ac.uk/xtf/view?docId=ark/32150_s173666459g.xml.

39 Back cover of Midgley, *The Owl of Minerva*. Italics mine.

Chapter 2: Human Nature and the Self

1 Midgley on the main themes or emphases of her work (Mary Midgley, Interview by Gregory S. McElwain, March 6, 2011, in Newcastle Upon Tyne, UK).

2 Midgley, *Beast and Man*, xiii–xiv. This approach had antecedents in empiricist theories of human nature, most notably John Locke's *tabula rasa* or "blank paper" account, in which we are essentially made by experience, not by nature (John Locke, *An Essay Concerning Human Understanding*, ed. Peter H. Nidditch [Oxford: Clarendon Press, 1979]).

3 Influential pioneers of behaviorism include John Watson, *Behaviorism* (Chicago, IL: University of Chicago Press, 1930); and B. F. Skinner, *Science and Human Behavior* (New York: The Free Press, [1953] 1965). The most influential sociobiologist was Edward O. Wilson, *Sociobiology: The New Synthesis* (Cambridge, MA: Belknap Press of Harvard University Press, [1975] 2000). Wilson was the most prominent representative of the more biologically determinant position in Midgley's early work though, over time, Richard Dawkins became a more consistent representative (Richard Dawkins, *The Selfish Gene* [Oxford: Oxford University Press, 1976]). Early sociobiology, under the influence of Wilson, made some extreme claims about the deterministic role that inherent traits play on human behavior. Some sociobiologists claimed that, like other animals, we are genetically programmed to behave in certain ways. As such, the most hardline defenders of this position insisted that putatively "external" factors, such as culture, have little to no impact on human behavior. The nature position, moreover, firmly grounded humans in the natural world. All organisms, including humans, operate under the same natural laws in the competition for survival. "Nature," in turn, was wielded to unmask some of our most cherished human features and challenge human exceptionalism. For instance, since genes truly control our behavior, seemingly altruistic acts were explained away as the activities of "selfish genes" seeking to propagate themselves. The underlying principles of self-preservation were invoked as clandestine motives in altruistic behavior, reducing all motives to egoism, calculating, or otherwise.

4 This move, especially in the 1970s–80s, was largely influenced by existentialism. While there are certain "givens" and conditions that do delimit human behavior to some extent, these are marginal in the face of the radically shaping potential of authentic human choice (Jean-Paul Sartre, "Existentialism Is a Humanism," in *The Continental Ethics Reader*, eds. Matthew Calarco and Peter Atterton [London: Routledge, 2003]).

5 Under the banner of the laws of human nature, Social Darwinists advocated for "survival of the fittest" (first coined by Herbert Spencer) in society as a natural moral law—only the strong (and selfish) survive (Herbert Spencer, *Principles of Biology, Volume I* [New York: Appleton, (1864) 1898], 530–31). Those who cannot keep up are naturally inferior.

6 Mary Midgley, *Can't We Make Moral Judgements?* (London: Bloomsbury, [1989] 2017), 87.

7 Ibid., 87. See also Midgley, *The Myths We Live By*, 152–53, for expansion on the moral elements of human nature. See also "Morality."

8 The study of human nature can be fraught with problems. Moreover, the sheer magnitude of the subject and how it should be methodologically approached is daunting. The key here, for Midgley, is that all available modes of inquiry and forms of knowledge are utilized (see "Science"). She discusses how differing viewpoints or perspectives help to construct the most complete knowledge. For instance, in studying a mountain, no single approach to the vastness of the mountain has the complete perspective (Midgley, *The Ethical Primate*, 43). In this example, overhead or distant viewpoints may generate helpful and dramatic images of the mountain, but more close up work is needed to understand the finer details. Likewise, a plurality of approaches rather than sweeping monistic approaches (such as sociobiology, for example) are needed to understand the intricacies of human nature, life, and behavior. A proper understanding of human nature requires that all modes of inquiry are present and, ideally, in conversation.

Midgley also laments the disciplinary fragmentation that makes this study challenging. Specialization has its merits in creating rich depths of knowledge, but often, especially in the case of the sciences, it leaves the specialist unpracticed in seeing the broader connections that make their work significant and worthwhile. This leads to specialists in all types of fields operating within their narrow silos without communicating with each other and connecting ideas. Her inquiry brings together everything from literature, poetry, and the arts to psychology, primatology, and genetics.

9 Midgley approaches the study of human nature in a naturalistic perspective grounded in a Darwinian understanding of human history, though she is generally cautious in her use of the term "naturalism" (Mary Midgley, Interview by Gregory S. McElwain, May 24, 2012, in Newcastle Upon Tyne, UK). This is because she resists the more hardline and reductive materialism characteristic of some forms. Her approach is more akin to what Ted Benton terms "non-reductive naturalism" (Ted Benton, *Natural Relations: Ecology, Animal Rights & Social Justice* [London: Verso Books, 1993]). Regardless of the term used, Midgley relies on the usual assumptions of naturalism regarding the workings and regularity of the natural world and the human place in it.

10 Mary Midgley, "The Concept of Beastliness: Philosophy, Ethics and Animal Behaviour," *Philosophy* 48, no. 184 (1973): 114.

11 Jean-Paul Sartre, "Existentialism Is a Humanism," 91.

12 Ibid., 133.

13 Starting especially in the 1970s, Midgley was keenly interested in and influenced by zoology and the field of animal studies. Ethology (the study of animal behavior) is an area that Midgley relied on in her early publications, along with such areas as primatology, human-animal studies, and virtually any area that carefully studies animals in a variety of contexts. She found in these studies a collection of scientific approaches that strove to understand animals as they really were rather than as fictionalized creations of the human psyche. Midgley also recognized that attentive descriptions of animal nature and behavior could provide the framework of human-animal comparison as a way of understanding human nature and behavior. She argues that, in a Darwinian context, it would be strange to believe that the concepts, methods, and comparisons utilized for other animals had no relevance or application to the human case (Midgley, *Beast and Man*, 14–15). Of course, comparisons and inferences should only be made in relevant cases appropriate to the relationship between species and circumstances (some early ethologists developed a habit of overstepping these bounds) (Ibid., 23 and 24). For an examination of the historical use and abuse of human-animal comparison, see David Cooper, *Animals and Misanthropy* (London: Routledge, 2018).

14 Midgley, "The Concept of Beastliness," 114.

15 This study is a more difficult undertaking than it may at first appear, partly because of the sheer taxonomic comprehensiveness of the term "animals," and partly because of how much conceptual and symbolic baggage has become tied up with the term. Animals are not a generic mass, but real creatures that differ in their habits, attitudes, behaviors, relations, and so on (see "Animals"). The term "animals" can sweepingly cover a whole spectrum of life from blue whales and other charismatic megafauna down to miniscule microorganisms such as tardigrades (not to mention all the variety of creatures from dogs to songbirds to insects that we often encounter on a daily basis) (Midgley, *The Myths We Live By*, 191). Species-barriers are also more convoluted than is often assumed, and many of the seemingly official boundaries that have been erected between animals tend to be driven by value judgments as much as scientific considerations (see "Animals") (Ibid., 196–97). The solidity of these barriers is, furthermore, often negotiable and adjusted ad hoc, often to suit the needs of different interests—for example, farmers, animal researchers, pet owners, stockyards, and so on. For this reason, Midgley highlights the complexity and challenges involved with the term "animals" and reminds us that our conversations about animals (which almost exclusively is in reference to *other* animals) must be particularized.

16 Ibid., 193–94.

17 See James Serpell, *In the Company of Animals* (Cambridge: Cambridge University Press, [1986] 1996).

18 Midgley, *Beast and Man*, 205.

19 Ibid., 39.

20 Midgley, "The Concept of Beastliness," 114.

21 Mary Midgley, "Heaven and Earth, an Awkward History," *Philosophy Now*, 2001–2002, 32 (Available online at https://philosophynow.org/issues/34/Heaven_and_Earth_An_Awkward_History).

22 Midgley, *The Ethical Primate*, 23–24. The spectrum of shared characteristics is further explored in "Animals."

23 Midgley, *Beast and Man*, 198 and 196.

24 Ibid., 153.

25 Ibid., 198.

26 Ibid., 17.

27 Ibid., 195.

28 See Ian Ground, "Minding Animals," in *Science and the Self: Animals, Evolution, and Ethics: Essays in Honour of Mary Midgley*, eds. Ian James Kidd and Liz McKinnell (London: Routledge, 2015), 129–48. The challenging issue Ground explores is to what extent these capacities are metaphysically significant.

29 Midgley, *Beast and Man*, 49–50. The term "instinct" is tied up with a range of words used to describe behavior, including "drive" (which deals with dynamic forces and intensity) and "program" (which deals with communication and coding) among others (Mary Midgley, *Heart and Mind: The Varieties of Moral Experience* [London: Routledge, (1981) 2003], 186–87). Midgley uses "instinct" over these other terms due to its less technical and more flexible usage in everyday language. "Instinct" can also capture the dynamic and communicative elements of "drive" and "program," respectively, and is connected to the concept of an "instigator," which can operate "both as a source of information on what to do, and as a force impelling us to do it" (Ibid., 187). Information in this context is a formula or specification that is responsive to its environment. Ultimately, "the nature of a species," she maintains, "consists in a certain range of powers and tendencies, a repertoire, inherited and forming a fairly firm characteristic pattern, though conditions after birth may vary the details quite a lot" (Midgley, "The Concept of Beastliness," 128). For Midgley, an ethologist studying these instincts and their connected behaviors is, consequently, "like a surveyor mapping a valley" (Ibid.).

30 Midgley, "The Concept of Beastliness," 125.

31 Midgley, *Beast and Man*, 51.

32 Midgley, "The Concept of Beastliness," 125–26.

33 Cats, for instance, are instinctively driven to hunt, but there is no single way to hunt. Hunting behaviors cover a wide "repertory of movements," some of which are learned and invented. Midgley, "The Concept of Beastliness," 126. Midgley, *Beast and Man*, 52. She notes that "open and closed instincts . . . are clearly not distinct kinds of things; they are extremes of a scale with many grades between" (Ibid.). And the explanatory power of a given motive, furthermore, varies depending on where it falls on the open-closed spectrum (Midgley, *Heart and Mind*, 192). Instincts cluster together in diverse ways and help, in context, to explain some types of behaviors in better ways than others. Humans and bowerbirds might gather blue flowers for analogous courtship-related purposes, though humans might gather flowers for any number of other reasons in accordance with a wider motivational structure (for play, for cultivating, for garnishing a salad, etc.). The rich diversity of human behavior and expression across cultures and throughout history is, in part, explicable through the profound range of open instincts that humans possess. Ibid., 202–03.

Human motivation is, of course, a complex topic and instincts can be accommodated in it in multiple ways. In one sense, Midgley's approach to instincts is akin to William James's "instinct theory of motivation" (William James, *The Principles of Psychology*,

Vol. 2 [New York: Dover, (1890) 1950]). However, James's theory seems to be more about fixed or closed instincts, whereas Midgley's approach includes flexible or open instincts. In this way, her approach allows for the variability of expression of instincts in behavior, the influence of experience, and so on. Furthermore, instincts can be seen as one among many influences on human motivation including rewards or incentives, needs, tension relief, utility, social conditioning, and so on. Instincts, as such, have dynamic interactions with numerous motivational variables. Though some may stress one particular theory of motivation as *the* explanation, motivation is best seen here in a comprehensive sense as having many influences and causes. Midgley recognizes a plurality of motives and argues that there is a general structure of motivation that underlies the various expressions of behavior.

34 Midgley, *Beast and Man*, 19.

35 Ibid., 28. "Culture" is a complicated term that has many contestations. Midgley uses this term mostly in an everyday sense. She typically uses it in reference to geographic distributions of human populations, recognizing the integrity of different regional heritages. Though the spatial and temporal boundaries of cultures are fuzzy, she occasionally uses this to recognize and distinguish between differing cultural backgrounds (e.g., in discussing cultural relativism—see "Morality"). In another sense, Midgley uses culture quite broadly as something universally shared and expressed by humans. "Human culture" here is something of a parallel to the broad and variable "human nature." This is the use that most often arises in discussing human nature. It is sometimes loosely interchanged with "society."

36 Midgley, *The Ethical Primate*, 175.

37 Midgley, *Wickedness*, 152. Understanding behavior also requires an account of such things as moods, affects, modes of perception, feelings, and thoughts. Envy, for instance, is not just a feeling, but rather a formula in which certain feelings combine with the thought that someone has something that you do not have (Ibid., 146–47). Feelings can move us, but in the case of pronounced motives such as envy, "the feeling alone is not the motive. Feelings are things which may happen to anybody, but to have a motive is to envisage a policy" (Ibid., 146).

38 Midgley, *Beast and Man*, 15. Midgley stresses elsewhere that "*there is not in animals any sign at all of a system like Freud's or Nietzsche's, whereby any major motive subsumes or rules the rest*" (Midgley, *Heart and Mind*, 201). Not only does this make little evolutionary sense (the obsessive animal dominated by one motive would likely not survive), but it does not provide enough explanatory power for much of our behavior (Midgley, *Beast and Man*, 161).

39 Mary Midgley, Interview by Gregory S. McElwain, October 24, 2017, in Newcastle Upon Tyne, UK. This plurality of motives can and does lead to conflict. Inevitably, such a complicated being with such an array of at times harmonious and at times clashing motives will run into inner conflict. The ordering of these conflicts is significant, and Midgley traces how this functions in Darwin's account of the origins and development of morality (see "Morality").

40 Midgley, *Heart and Mind*, 204.

41 Ibid., 7–12.

42 Ibid., 8. This, again, is informed by the contrasting influences of dualistic religion and Sartrean existentialism (as opposed to Simon de Beauvoir's later vision of will interacting with "circumstance" and other factors) (Simone de Beauvoir, *Force of Circumstance, Vol. I: After the War, 1944-1952; Vol. 2: Hard Times, 1952-1962*, translated by Richard Howard [New York: Paragon House, 1992]).

43 Midgley, *Wickedness*, 96.

44 Ibid.

45 Ibid., 96–97.

46 Midgley, *Beast and Man*, 61.

47 Midgley, *Heart and Mind*, 40.

48 Midgley, *Beast and Man*, 315–16.

49 Midgley, *Are You an Illusion?* 104.

50 René Descartes, *Meditations on First Philosophy: With Selections from the Objections and Replies*, trans. John Cottingham (Cambridge: Cambridge University Press, [1986] 2017). Midgley seems to endorse the familiar critiques of dualism and the mind as substance—the inability of mind to interact with matter, the challenges of evolutionarily explaining the rise of a substance other than matter in naturalistic terms, and so on (Midgley, *Science and Poetry*, 153.).

51 This idea of the self as an illusion is, aptly, Midgley's main target in *Are You an Illusion?* Much of the illusion position, as she argues in 2014, is more about metaphysics than science per se, which can happen when much of the scientific community is not trained in philosophy or is not aware of their own philosophical stances (or confuse philosophical positions with scientific ones). Consequently, metaphysical ideas are often dressed as scientific ideas. "This thought," she writes, "about the self, dismisses it as unreal. That, of course, is bad metaphysics in a pretty obvious sort of way. People should know that metaphysics matters and pay a bit of attention to it" (Midgley, Interview by Gregory S. McElwain, June 11, 2014). The self as an illusion is, rather, one imaginative way of seeing the self among many (Midgley, *Are You an Illusion?* 7) (see "Science").

Some materialists argue that consciousness is an epiphenomenon. Here, mental properties are by-products of physical states in a monocausal direction. Like a train whistle blowing off steam from the engine, consciousness is a response to physical processes and stimuli but cannot supervene on them (T. H. Huxley, "On the Hypothesis That Animals Are Automata, and Its History," *Fortnightly Review* 95 (1874): 555–80). This fatalistic account leaves us with no free will or ability to influence the world. Midgley discusses this in 2014: "This is . . . the notion that physical matter determines everything. I mean, it's quite an interesting thing that I mention in [*Are You an Illusion?*]. Take this metaphor of T. H. Huxley's steam whistle. Now, if that were right, the thought would always be produced by the brain cells. So, when Einstein is doing his 'stuff,' how do the brain cells know that $E=MC^2$? How did they get there first? It isn't at all plausible, is it? . . . It seems to me another superstition, that physical matter is always at the causal end. That the thought is counterfeit . . . [is a] superstition—a quite strong conviction—that physical matter has always got to be the cause" (Midgley, Interview by Gregory S. McElwain, June 11, 2014).

52 Midgley, *Science and Poetry*, 92–93.

53 Ibid., 15. This aligns largely with Ryle's critique of dualistic approaches to mind, in which he argues that reference to the substance of mind is a "category mistake" (Gilbert Ryle, *The Concept of Mind: 60th Anniversary Edition* [London: Routledge (1949) 2009]).

54 Since much of the Western philosophical tradition has been preoccupied with elevating reason above many things—nature, other animals, bodies, emotions—a large portion of Midgley's expositions on reason has to do with bringing it back down to earth, so to speak. She reiterates that reason, as an evolved capacity, has developed and proliferated in a way that is grounded in our history on this planet. As such, it helps us sort and evaluate means and ends, make decisions, and organize ourselves and our inner conflicts in the ways necessary for social and moral creatures (Midgley, *Beast and Man*, 248). Midgley is most concerned with the tradition's overemphasis on the *rule* of reason. She presents the model of reason in the tradition of Plato, the Stoics, Descartes, Spinoza, and Kant (Ibid., 250). Much in the way that colonists prided themselves in their rule over "foreign" subjects, this type of rationalism subjugates the passions as something "other" to be controlled at the risk of chaos. While it is important to value reason in human life, this colonial model has become hypertrophied and must be reimagined in the more sensible context of the whole person. Midgley draws on the traditions of Bishop Butler and Aristotle, to an extent, in conceiving of reason as arising out of and completing the whole person (Ibid.). Reason, in this picture, is less of a dominating force and more of a cooperative organizer or facilitator of the whole. Our perceptions and feelings feed into this dynamic network and contribute to our thoughts and, eventually, aims (Midgley, *Heart and Mind*, 4, and Mary Midgley, *The Solitary Self: Darwin and the Selfish Gene* [Durham, NC: Acumen, 2010], 64–65). Likewise, our thoughts impact our perceptions and feelings in a feedback cycle in which all of these elements are inextricably bound together.

Midgley argues that Kant too closely associates feeling with the overly dramatic and grandiose passions of the German Romantics and, in turn, reacts against the all-consuming and controlling feelings associated with such things as nationalism. As such, Kant quite understandably thinks reason should be separated and protected from the influence and interference of such dramatic and, at times, dangerous passions. In this drama, reason must maintain its central role as the great ruler of the passions and as the distinctive mark of humanity (and the moral community). However, in assuming a strict separation between reason and emotion, Kant overlooks the feelings that move us to reason and moral philosophy in the first place. He too sharply separates out the formal from the emotional elements of such things as respect, for instance (Midgley, *Heart and Mind*, 106 and 111). She writes elsewhere that respect "positively calls for coherent reasoning, emerging sometimes into rules. But this intellectual aspect does not conflict with the emotional one. It complements it" (Midgley, *The Ethical Primate*, 151–52). She later comments: "If you didn't have any motives, if you didn't have feelings, reason alone can't get you going, can it? It's a formal process of balancing" (Midgley, Interview by Gregory S. McElwain, May 29, 2013).

55 Midgley, *Heart and Mind*, 102–03.

56 It is this dynamic process of analyzing and articulating both thoughts and feelings that gives rise to our standards, judgments, and priority systems (see "Morality") (Midgley, *The Myths We Live By*, 150).

57 Midgley, *Science as Salvation*, 129.

58 Midgley, *Science and Poetry*, 150. Midgley also writes that "feelings, to be effective, must take shape as thoughts, and thoughts, to be effective, must be powered by suitable feeling" (Midgley, *Heart and Mind*, 4). To assume that we rely on emotion alone—for instance, in morality, as emotivist theories maintain— "ignores reflection, rumination, contemplation, brooding, worrying, dreaming, reminiscing, speculating, considering and imagining. In particular, it ignores deliberate redirecting of attention by which we can, if we please, gradually transform our feelings" (Midgley, *The Solitary Self*, 75).

59 Midgley, *The Ethical Primate*, 15.

60 These distinctions are laid out in Midgley, *The Ethical Primate*, 123–24.

61 Thomas Hobbes, *Leviathan: With Selected Variants from the Latin Edition of 1668*, ed. E. M. Curley (Indianapolis, IN: Hackett, 1994), 76.

62 John Locke, *Two Treatises of Government*, ed. P. Laslett (Cambridge: Cambridge University Press, [1690] 1967); Jean-Jacques Rousseau, *'The Social Contract' and Other Later Political Writings*, trans. Victor Gourevitch (Cambridge: Cambridge University Press, 1997).

63 Midgley, *The Solitary Self*, 121. "Freedom for" and "freedom from" parallel Isaiah Berlin's account of positive and negative liberty (Isaiah Berlin, *Four Essays on Liberty* [Oxford: Oxford University Press, 1990]).

64 Mary Midgley, "Philosophical Plumbing," 144.

65 Mary Midgley, *Evolution as a Religion: Strange Hopes and Stranger Fears* (London: Routledge, [1985] 2002), 136–37.

66 Midgley, *The Ethical Primate*, 119.

67 Midgley, "Philosophical Plumbing," 145, italics mine.

68 Ibid., 146–47.

69 Midgley, *The Solitary Self*, 140.

70 Midgley, *Science and Poetry*, 20.

71 Midgley, *Can't We Make Moral Judgements?* 111.

72 Mary Midgley, "The Objection to Systematic Humbug," *Philosophy* 53, no. 204 (1978): 169.

73 Midgley, *Heart and Mind*, xvi.

Chapter 3: Morality and Wholeness

1 Midgley, Interview by Gregory S. McElwain, May 23, 2012.

2 Mary Midgley, *Heart and Mind*, 123. This diverges from positions that emphasize content, such as happiness or interests. For Midgley, happiness is not the end of morality, but rather a part of it. This is because happiness is something serious or important. Midgley here utilizes part of Philippa Foot's argument that content along with seriousness distinguish moral issues (Philippa Foot, "The Philosopher's Defence of Morality," *Philosophy* 27, no. 103 [1952]: 311–28). Midgley reiterates these aspects of morality in 2017: "The chief reason, I suppose, for saying that something is a moral distinction is that you're denying it's a trifling issue. It's serious. And something being serious is a perfectly clear sort of thought" (Midgley, Interview by Gregory S. McElwain, October 24, 2017).

3 Mary Midgley, "Is 'Moral' a Dirty Word?" *Philosophy* 47, no. 181 (1972): 221.

4 Ibid., 223.

5 Ibid.

6 Ibid.

7 Ibid., 226.

8 Midgley, Interview by Gregory S. McElwain, June 11, 2014.

9 Midgley, *Beast and Man*, 166.

10 Ibid., 160. Good moral philosophy, for Midgley, takes cultivation, practice, and a certain sense of the whole. It requires not only meticulous detailed thinking—lawyer-like doggedness and attention to detail—but also poet-like attention to competing visions of life. This is different than finding straightforward solutions to problems or passing an exam. She reflects on this in 2016: "What happens is someone comes straight from school and they think, 'Here's this subject. What are the definitions?' They go straight into it. Aristotle said somewhere that you can't do moral philosophy until you're over 40. I mean, he had a point. You don't have to be 40, but you should be coming to it from life and not from seeing it as one of a number of subjects that you might pass exams in" (Mary Midgley, Interview by Gregory S. McElwain, May 30, 2016, in Newcastle Upon Tyne, UK).

11 Mary Midgley, *Wisdom, Information and Wonder: What Is Knowledge For?* (London: Routledge, 1989), 119.

12 Midgley, *Heart and Mind*, 17.

13 Mary Midgley, "Biotechnology and Monstrosity: Why We Should Pay Attention to the 'Yuk Factor,'" *The Hastings Center Report* 30, no. 5 (2000): 10. This is important in thinking of our connections not just to each other but also to animals and the wider world (see "Animals" and "Nature").

14 Charles Darwin, *The Descent of Man* (Princeton, NJ: Princeton University Press, [1871] 1981).

15 Midgley, *The Solitary Self*, 31. There is a deep-seated dispute here about egoism and altruism. Midgley takes aim at the sociobiologist reduction of all altruistic behavior to egoism. There are many facets to this debate, but she seeks to undermine the idea that sociobiology unmasks or debunks altruistic behavior. Her argument throughout is that the affections and bonds that motivate humans are deeply rooted in our social instincts. Given the strengths of the bonds, especially parent-child bonds, that are needed to ensure survival and development, it makes sense that these strong feelings and affections can naturally expand and extend to others (Midgley, *Beast and Man*, 130–31). Though these motives may enhance fitness (and that of others close to us), it does not explain away that we do behave altruistically and have good will toward others (Ibid., 121).

16 This broadly reflects Hume's account of the moral sentiments and the rise of morality, since Darwin was influenced by Hume. However, Midgley argues that Darwin's account diverges by more fully integrating rather than rigidly dividing reason and emotion. David Hume, *A Treatise of Human Nature*, eds. David Fate Norton and Mary J. Norton (Oxford: Clarendon Press, [1738] 2000).

17 Midgley, *The Ethical Primate*, 6 and 159.

18 Ibid., 140 and 145.

19 Midgley, *The Solitary Self*, 61.

20 Darwin's full quote: "Man, from the activity of his mental faculties, cannot avoid reflection. . . . Any animal whatever, endowed with well-marked social instincts, would inevitably acquire a moral sense or conscience, as soon as its intellectual powers had become as well-developed, or anything like as well-developed, as in man" (Darwin, *The Descent of Man*, 72).

21 Midgley, *Wickedness*, 190.

22 Midgley, *The Solitary Self*, 80.

23 Midgley, *Wickedness*, 198–99.

24 Midgley, *The Ethical Primate*, 144.

25 Ibid., 181–82.

26 Ibid., 181.

27 Midgley, Interview by Gregory S. McElwain, May 24, 2012.

28 Midgley, *Beast and Man*, 175–76.

29 Ibid., xli–xlii. There are some pitfalls that must be avoided in connecting human nature and morality, namely, in making the move from facts to values in moral reasoning. "Crude and reductive" forms of naturalism made this mistake by claiming that the "natural law" of survival of the fittest should guide social ethics. Social Darwinists promoted this view—and distorted Darwin's views on morality in the process—arguing that nature *is* characterized by cutthroat competition and that human society *should* follow suit. The strong survive (and *should* survive) and the weak perish (and *should* perish). This type of position elicited anti-naturalist criticism, which argued that these positions made wild jumps between natural facts and values, assuming that *evolved* or *natural* is *good* without further justification, elaboration, conceptual analysis, and moral argumentation (not to mention, incidentally, that the facts are likewise dubious and contested in this case). They thus violated both Hume's is/ought-fact/value dichotomy and Moore's naturalistic fallacy. Moore argued that what is "good" cannot be identified with any natural property; "good," rather, is a non-natural property that is intuited (G. E. Moore, *Principia Ethica* [Cambridge: Cambridge University Press (1903) 1948]).

 Midgley is likewise critical of reductionist approaches to morality that conflate facts and values. She argues that the problems with this sort of naturalism primarily stem from muddled thinking and poor argumentation. She also investigates the conceptual relationship between facts and values, which is more complicated. Part of the challenge in dealing with facts and values is that there is less of a divide between certain facts and values. "Facts" and "values," in other words, are often more interrelated than supposed. Facts are too often seen as straightforward and unproblematic data. Yet, she argues, they are never truly neutral, primarily because "we are never neutral even in what we 'see'— we must always select, interpret, and classify" (Midgley, *Beast and Man*, 170.) This is true in everyday life and in the scientific sense. There is a world out there, but we do not approach it as raw or brute data (see "Science"). As such, "anything we can think about at all has been already shaped by our concepts in the process of perception. And the data of any serious moral problem always incorporates quite complex conceptual schemes" (Mary Midgley, "The Flight From Blame," *Philosophy* 62, no. 241 [1987]: 277). In this way, concepts and values are incorporated and built into facts over time, so much so that "particular puzzles can be so fully solved that they are forgotten, and later generations see their solutions as simply part of 'the facts'" (Ibid.). This, of course, does not mean that facts *are* values, or that they clearly outline values, as some have claimed (E. O. Wilson and B. F. Skinner are examples here. Midgley, *The Ethical Primate*, 72–74). It simply means that a realist view of facts as straightforward and unproblematic does not fully consider their complexity. She discusses the notion of "importance" and how it lights up some of the interconnection and complexity of facts and values: "The facts are supposed to be self-contained bunch, aren't they? But if you begin to think about a concept like importance or danger, and ask if this is a factual concept or a value one, you are rather quickly in trouble, aren't you? Because to say that something is important is, among other things, to say that it is connected factually with a lot of other things which produce these consequences or the like. I think importance is a very crucial one. Danger is similar, you see. To say that something is dangerous is both to say something factual about the effects it's going to have and to take a stand for or against those consequences" (Midgley, Interview by Gregory S. McElwain, May 29, 2013). There is a distinction in concept between the judgments one makes about the world and the world about which those judgments are made. But they are likely more entangled than some assume.

Benjamin Lipscomb very nicely summarizes how Midgley built on the influential work and critique of her Quartet colleagues on some of these conceptual issues related to human goods: "The key points, for Midgley's purposes, were these: 1. There is no need to accept an anti-naturalist account of moral language and judgment. Indeed, important features of moral thought and discourse are mystified if one does not understand them as bearing an essential, 'internal relation' to human needs (Foot, 'Moral Beliefs'). 2. Aristotle showed us, in outline, how to construe this relation. Our good is bound up with the traits and circumstances creatures like us need in order to flourish in a characteristically human life (Anscombe, 'Modern Moral Philosophy'). 3. These points hold even if we limit our investigation to highly abstract evaluative words/concepts like 'good,' but they hold more manifestly if we consider the richer body of words/concepts people use in judgments of character . . . (Anscombe and Foot). 4. There can be in any case no neutral analysis of moral language and moral judgment (Murdoch, 'Metaphysics and Ethics' *et passim*)." Lipscomb, "Slipping Out Over the Wall," 215.

30 Midgley, *Heart and Mind*, 54–55.

31 Values here are not so much created as emphasized: "We've got the values from the start, we just decide which of them to emphasize" (Midgley, Interview by Gregory S. McElwain, June 11, 2014).

32 Midgley, *The Myths We Live By*, 137.

33 Midgley, *Can't We Make Moral Judgements?* viii.

34 Midgley, Interview by Gregory S. McElwain, May 24, 2012.

35 Midgley, Interview by Gregory S. McElwain, May 29, 2013.

36 Ibid.

37 Midgley, *The Myths We Live By*, 223–24.

38 Midgley, Interview by Gregory S. McElwain, May 29, 2013.

39 Midgley, Interview by Gregory S. McElwain, October 24, 2017. Furthermore, standards such as consistency alone do not necessarily make a good theory: "Theory has to take its place along with everything else. Just because everything is consistent with whatever you previously decided, isn't enough to show that it's right" (Midgley, Interview by Gregory S. McElwain, May 29, 2013).

40 Ibid.

41 Midgley, *Utopias, Dolphins and Computers*, 130–31.

42 Midgley, *Beast and Man*, 160.

43 Midgley, *Utopias, Dolphins and Computers*, 131.

44 Midgley, *The Owl of Minerva*, 65.

45 Midgley, *Wickedness*, 26.

46 Midgley, Interview by Gregory S. McElwain, May 29, 2013.

47 Midgley, Interview by Gregory S. McElwain, June 11, 2014. She expands: "It seems to me that once we believe that humans are social animals it will follow that it matters how they treat each other. . . . It is a description of how human life works. . . . So morality is not, as I said, something separate that's imposed—it grows out of it."

48 Midgley, *Can't We Make Moral Judgements?* 53.

49 Ibid.

50 Ibid., 34.

51 Midgley, *The Solitary Self*, 64–65.

52 Ibid., 64.

53 Furthermore, it is difficult to imagine how certain moral issues make sense outside of our deeply social context. For example, Midgley explores the *relational* elements of telling the truth: "Now we are not only talking about cases where the duty to tell the truth conflicts with some other duty. We are talking about the sort of truthfulness which we should be looking for in ordinary life, truth that we owe each other, so to speak. The kinds of duties that are involved with being a human being . . . are not something optional. This is because we are in relation with others. A relation is something real in the world. We have to carry it on in certain sorts of ways" (Midgley, Interview by Gregory S. McElwain, October 24, 2017).

54 Midgley, *The Solitary Self*, 125.

55 Midgley, *Can't We Make Moral Judgements?* 21–22.

56 Ibid., 27. "Dogmatic skepticism," as Midgley calls it, tends to rule out the possibility of moral judgments, claiming we are unable to ever truly know if an action is wrong (Midgley, *Can't We Make Moral Judgements?* 11). This coincides with an argument for "freedom from interference," arguing that we should not judge or interfere with others since we can never *know* what is truly right or wrong. She argues that this argument typically assumes that freedom is good or the ultimate value, which is itself a moral judgment. Moreover, freedom typically requires positive argumentation—for instance, for the value of independent thought or self-reliance, and so on, which is often absent in such positions. This type of position, she argues, tends to ultimately promote its own particular set of unquestioned values, many of which cannot be supported in ways that are not self-defeating to the project of moral skepticism. This "selective skepticism" values such virtues as incredulousness and denigrates others such as credulity (Midgley, *Evolution as a Religion*, 125–26).

57 Midgley, *Can't We Make Moral Judgements?* 28–29. Midgley separates "moral isolationism" from subjectivism, which often get lumped together as an overarching relativism. She characterizes subjectivism as a set of values and claims surrounding interpersonal skepticism and judgments. There are, of course, important elements of subjectivism that should be recognized— subjectivity and difference, for instance, along with personal choice and freedom—but these should not be elevated to the point of isolationism that separates people in the way that cultural relativism isolates cultures. Rather, we are interconnected and share, contribute to, and draw from common sources of values in our moral systems (Ibid., 7). Standards are liable to vary widely, but this does not disconnect them. As social creatures, we do not create these values through some act of the will, but acquire them in this context. Here we can sort through and make cases for why certain priorities and values should be emphasized more than others, but this is a matter of *emphasis* rather than morality by private decision.

58 Midgley, *Heart and Mind*, 80.

59 Midgley, *Can't We Make Moral Judgements?* 149.

60 This example is explored in Midgley, *Heart and Mind*, 81–85.

61 Midgley, *Can't We Make Moral Judgements?* 76–77.

62 Ibid., 91.

63 Ibid., 85.

64 Midgley, *Animals and Why They Matter* (Athens, GA: University of Georgia Press, 1983), 109.

65 Mary Warnock argues that this is where certain elements of consensus morality make sense (Mary Warnock, "Is There Such a Thing as Consensus Morality?" in *Science and the Self: Animals, Evolution, and Ethics: Essays in Honour of Mary Midgley*, eds. Ian James Kidd and Liz McKinnell [London: Routledge, 2015], 85–99).

66 The history of moral reform can quite reasonably be seen as a history of internal critique, or the process through which groups look inward and interrogate, criticize, and augment their value systems. Midgley notes that much of Western culture has been preoccupied with this very project (Midgley, *Beast and Man*, 283). Moreover, internal critique is not purely internal, as many of the sources and portions of morality in any cultural system are typically integrated (or appropriated) over time from other cultures and contexts.

67 Midgley, *Can't We Make Moral Judgements?* 140.

68 Though, at times, we have to come to terms with the fact that they do not. Midgley discusses this in 2017: "There's no general recipe for changing other people's opinions. It's something we often want to do, and we try various methods. What we shouldn't do is shoot them or shut them up. We are naturally inclined to hope that other people will agree, but we have to be more or less realistic and truthful about this. They often don't" (Midgley, Interview by Gregory S. McElwain, October 24, 2017).

69 Midgley, *Science and Poetry*, 230–32.

Chapter 4: Animals and Why They Matter

1 Midgley, Interview by Gregory S. McElwain, March 6, 2011.

2 Midgley, *The Owl of Minerva*, 6.

3 Midgley, *Utopias, Dolphins and Computers*, 117.

4 Genesis 1:26–28. Scripture quotations are from the *New Revised Standard Version Bible* copyright © 1989, 1995 National Council of the Churches of Christ in the United States.

5 Midgley, *Beast and Man*, 153.

6 Midgley, *Animals and Why They Matter*, 17.

7 René Descartes, *Discourse on the Method*, trans. Donald A. Cress (Indianapolis, IN: Hackett Publishing, [1637] 1998), and Baruch Spinoza, *Ethics*, trans. E. M. Curley (Princeton, NJ: Penguin, [1677] 2005), Part IV, proposition 37, note 1.

8 Spinoza, *Ethics*, Part IV, proposition 37, note 1.

9 Ibid., 14–17.

10 Ibid., 17.

11 Ibid., 13.

12 Ibid.

13 Certain studies show "an association between animal abuse and a variety of antisocial behaviors" (Arnold Arluke, Jack Levin, Carter Luke, and Frank Ascione, "The Relationship of Animal Abuse to Violence and Other Forms of Antisocial Behavior," *Journal of Interpersonal Violence* 14, no. 9 [1999]: 963–75).

14 Voltaire, *The Philosophical Dictionary*, trans. H. I. Woolf (New York: A. A. Knopf, 1924), "Animals" (Available online at https://history .hanover.edu/texts/voltaire/volanima.html).

15 Jeremy Bentham, *Introduction to the Principles of Morals and Legislation* (New York: Hafner Publishing Co., [1789] 1948).

16 Midgley, *Animals and Why They Matter*, 47.

17 Ibid., 48. Hume's exclusion of animals from justice resulted in the course of his efforts to narrow justice to rules regarding property, as a "secondary" component of morality (Ibid., 50, and Midgley, *Evolution as a Religion*, 179). Like Hume, John Rawls's more recent and highly influential conception of justice is one that pushes animals out of the equation (John Rawls, *A Theory of Justice* [Cambridge, MA: Belknap Press, (1971) 2005]). However, though Rawls focuses on justice to the extent that it neglects other virtues (and much of the background for justice itself), he implores his readers to consider the variety of issues he omits, including animals and nonhuman nature. For both Hume and Rawls, absence from a narrow, legal sense of justice does not mean exclusion from morality.

18 Midgley, *Animals and Why They Matter*, 52. "So far as animals are concerned, we have no direct duties. Animals are not self-conscious, and are there merely as a means to an end. That end is

man. . . . If a man shoots his dog because the animal is no longer
capable of service, *he does not fail in his duty to the dog, for the
dog cannot judge*, but his act is inhuman and damages in himself
that humanity which it is his duty to show towards mankind. . . .
He who is cruel to animals becomes hard also in his dealings with
men" (Immanuel Kant, *Lectures on Ethics*, trans. Louis Infield
[Indianapolis, IN: Hackett Publishing, 1980]). These vague near-
duties toward animals appear to be binding in some way, even if
they can be overridden by other duties. Midgley argues that the
Kantian preoccupation with limiting the scope of duties highlights
the tension between the need to acknowledge that animals
should be properly treated and the inflexibility of these restrictive
categories of moral worth. She finds R. M. Hare more helpful
here, agreeing that we do not have the same duties to all species,
including humans, and that distinguishing types of duties toward
certain needs is a central concern of morality (R. M. Hare, *Freedom
and Reason* [Oxford: Oxford University Press, 1963]).

19 Midgley, *Animals and Why They Matter*, 61.

20 Ibid., 83.

21 Ibid., 54. Midgley, paraphrasing from R. G. Frey, *Interests and
Rights: The Case against Animals* (Oxford: Clarendon Press,
1980). She argues that Frey's claims are verbal and based on the
terms used rather than on any reasoning or reflection on actual
animal behavior. Further, the source of his terms, according to
Midgley, is in the English law relating to property rights, which
can only address humans. These property laws are very particular
and based in the principle of mortmain, a law devised to prevent
deceased people from owning property, which cannot be seen as
an authoritative source for relations with animals. Others, such
as Michael Leahy, follow suit in attempting to undercut and clarify
the terms and concepts that are used in reference to animals and,
in turn, animal liberation (Michael P. T. Leahy, *Against Liberation:
Putting Animals in Perspective* [London: Routledge (1991) 1994]).

22 Midgley, *Animals and Why They Matter*, 64.

23 Ibid., 83.

24 Garrett Hardin famously uses the lifeboat analogy against other
humans in consideration of the sustainability of the earth's carrying
capacity (Garrett Hardin, "Lifeboat Ethics: The Case against Helping
the Poor," *Psychology Today* September (1974): 800–12).

25 Midgley, *Animals and Why They Matter*, 20.

26 Singer claims that the capacity to suffer, which is determined by
the structure of the nervous system, stops "somewhere between

a shrimp and an oyster" (Peter Singer, *Animal Liberation*, 2nd ed. [New York: New York Review of Books, (1976) 1990], 174). He also argues that animals beyond this line should still be given the benefit of the doubt.

27 Tom Regan, *The Case for Animal Rights* (Berkeley, CA: University of California Press, 1983), 243.

28 For other influential approaches, see, for instance, Carol J. Adams and Josephine Donovan, eds., *Animals and Women: Feminist Theoretical Explorations* (Durham, NC: Duke University Press, 1995); Stephen Clark, *The Moral Status of Animals* (Oxford: Clarendon Press, 1977); Andrew Linzey, *Why Animal Suffering Matters* (Oxford University Press, 2009); Clare Palmer, *Animal Ethics in Context* (New York: Columbia University Press, 2010); James Rachels, *Created from Animals: The Moral Implications of Darwinism* (Oxford: Oxford University Press, 1990).

29 Midgley, *Animals and Why They Matter*, 66.

30 Midgley, *Utopias, Dolphins and Computers*, 130.

31 Midgley, *Evolution as a Religion*, 80.

32 Ibid., 92.

33 Ibid.

34 Ibid., 11.

35 See, for instance, Barbara King, *Personalities on the Plate: The Lives and Minds of Animals We Eat* (Chicago, IL: University of Chicago Press, 2017); Frans de Waal, *Are We Smart Enough to Know How Smart Animals Are?* (New York: W.W. Norton & Company, 2016); Nathan H. Lents, *Not So Different: Finding Human Nature in Animals* (New York: Columbia University Press, 2016); and the ethological works of Marc Bekoff (e.g., *Minding Animals: Awareness, Emotions, and Heart* [Oxford: Oxford University Press, 2003]), et al.

36 Midgley, *Are You an Illusion?* 85–86.

37 Paul Taylor, "The Ethics of Respect for Nature," *Environmental Ethics* 3, no. 3 (1981): 197–218; Regan, *The Case for Animal Rights*.

38 Midgley, *Beast and Man*, 335.

39 Frans de Waal, in particular, has argued the case for use of "anthropomorphic" language in relation to chimpanzees. See Frans de Waal, *Mama's Last Hug: Animal Emotions and What They Tell Us about Ourselves* (New York: W.W. Norton & Company, 2019).

40 Midgley, *Animals and Why They Matter*, 115.

41 Ibid., 128.

42 Ibid., 124. Midgley points out that the term "anthropomorphism" was originally invented for talk about God. Here, it was meant to show that God is beyond human life and characteristics. Thus, any description of God would, in turn, be a mirror of humanity rather than an accurate description of God (Ibid., 127). More recently, she argues, it has been transferred, virtually unchanged, onto animals.

43 Midgley, *The Ethical Primate*, 23–24. Regarding the development or morality, she writes that other animals "do not have anything like our power of dealing with it by standing back from their various motives, by taking the point of view of the whole, and trying to make some kind of balanced decision. They have other distinctive powers, but not this one" (Ibid.).

44 There have been some discrepancies in the exact genetic match between humans and chimpanzees and bonobos, ranging from 95 to 99%.

45 Midgley, *Beast and Man*, 208. Midgley recounts how the chimpanzee Ameslan seemed to lack the ability to speak not because of lack of intellect, but rather because of lack of the appropriate physiological connections between the brain and the larynx that exist in humans. However, inability to verbally express ideas, which, as she notes, is typical of human expression at times, does not invalidate other forms of communication or even thought processes. She expands on this in 2012: "I think people are still extraordinarily stuck with the thought that you can't *think* anything that you can't say. Isn't that very odd? Every day of our lives we are trying to say things that we can't say" (Midgley, Interview by Gregory S. McElwain, May 23, 2012). Rather, both verbal and nonverbal communications are intertwined components of the same process, and many animals possess sophisticated forms of communication, both within and between species. (Dogs communicate with each other and with humans with ease.)

46 The notion of "speciesism" (in certain versions, at least—she notes that Peter Singer's use of the term is more nuanced than others) maintains that preference for one's species is morally comparable to racism. She argues that speciesism and racism may possess similar superficial characteristics, but the deeper parallel falls apart at the species boundary. Racism is, of course, a horrible evil. However, she is concerned with how much the *concept* of racism parallels the *concept* of speciesism. The concept of speciesism, for Midgley, has value in drawing attention to oppression and injustice, but is not an adequate parallel to racism. "Race" in humans is an artificial distinction, whereas species is a biological distinction. Midgley maintains that the concept of racism typically involves a grouping of several problems, including group selfishness, asymmetrical power relations and hierarchies, and the drawing of trivial distinctions. Midgley argues that group selfishness and power dynamics pertain to speciesism, but the triviality of the distinctions drawn does not. This is mainly where the parallel falls apart. The charge of speciesism, in the end, might be more of a placeholder for the unstated problems of injustice, bias, self-interest, exploitation, and oppression (Midgley, *Animals and Why They Matter*, 101).

47 Ibid., 99.

48 Raymond Williams, *Problems in Materialism and Culture: Selected Essays* (London: Verso Classics, 1997). Williams talks of the wild contrast between a butterfly and a cyclone as conflicting images of the harmony and ruthlessness of nature.

49 Midgley, *Utopias, Dolphins and Computers*, 116.

50 Midgley, *Animals and Why They Matter*, 111 and 112.

51 Midgley, *Utopias, Dolphins and Computers*, 111–12.

52 Midgley, *Animals and Why They Matter*, 114.

53 Ibid., 114 and 116.

54 Ibid., 118.

55 Ibid., 119.

56 Ibid., 118.

57 Ibid., 119.

58 Ibid., 43.

59 Midgley, *Beast and Man*, 212.

60 Ibid., 22–23.

61 Ibid., 30.

62 Ibid.

63 Ibid., 26. Midgley intentionally utilizes the term "claims" in order to avoid some of the more technical quagmires that arise around such loaded terms as "duty" or "rights." Claims can likewise become caught up in technical and verbal quarrels, but she prefers to use it as a "working term" in the everyday sense because "rights are final and claims are not" (Midgley, Interview by Gregory S. McElwain, May 23, 2012). Midgley's typical stance on technical terms or categories is that they should be used insofar as they are helpful, especially in the everyday sense. If certain terms prove unsatisfactory or too weighed down with conceptual baggage, other language may be necessary. Claims, further, exist in varying degrees, independent of a being's or entity's ability to express the need for care and concern. In this sense, claims are not necessarily technical requirements imposed on us by rights-bearers (though they may share many of the same qualities).

64 Midgley, *Animals and Why They Matter*, 30.

65 Midgley, Interview by Gregory S. McElwain, May 23, 2012.

66 Others have highlighted the complex and contested boundaries of domestic and wild animals, for example: Grace Clement, "The Ethic of Care and the Problem of Wild Animals," *Between the Species*, 3 (2003); Clare Palmer, "What (If Anything) Do We Owe to Wild Animals?" *Between the Species*, 16, no. 1 (2013): 15–38; Anna L. Peterson, *Being Animal: Beasts and Boundaries in Nature Ethics* (New York: Columbia University Press, 2013).

67 John Hadley recognizes this in John Hadley, "Critique of Callicott's Biosocial Moral Theory," *Ethics & the Environment* 12, no. 1 (2007): 67–78.

68 Midgley, Interview by Gregory S. McElwain, March 6, 2011.

69 Ibid. She also suggests that the "mixed community" is a conceptual tool and that we should not get too locked into its use: "I use words as and when they seem appropriate. And sometimes one is talking at a very general level, and the abstractions are very important. And sometimes you get nearer to things and other words are needed."

70 "The point about wolves, for instance, is that you don't say: 'this [animal] is domestic; this one is wild—so I won't get involved in it, as it were.'" (Midgley, Interview by Gregory S. McElwain, May 23, 2012).

71 This divergence between animal and environmental approaches was most forcefully advocated by J. Baird Callicott ("Animal Liberation: A triangular Affair," *Environmental Ethics* 2, no 4. [1980]: 311–38; and Mark Sagoff, "Animal Liberation and Environmental Ethics: Bad Marriage, Quick Divorce," *Osgoode Hall Law Journal* 22, no. 2 [1984]: 297–307). See Gregory S. McElwain, "Midgley at the Intersection of Animal and Environmental Ethics," *Les Ateliers de l'Éthique / the Ethics Forum* 13, no. 1 (2018): 143–58, for an overview of this divergence and later efforts toward convergence with the help of Midgley's work.

72 Midgley, *Animals and Why They Matter*, 92.

73 Ibid., 93.

74 Taylor, "The Ethics of Respect for Nature," and Holmes Rolston III, *Environmental Ethics: Duties to and Values in The Natural World* (Philadelphia, PA: Temple University Press, 1989).

75 Ibid., 120.

76 See Callicott, "A Triangular Affair."

77 Mary Midgley, "Beasts Versus the Biosphere?" *Environmental Values* 1, no. 1 (1992): 114.

78 Ibid.

79 Midgley, Interview by Gregory S. McElwain, May 23, 2012.

80 Midgley, *Animals and Why They Matter*, 91.

81 Midgley, "Beasts Versus the Biosphere?" 117.

82 Ibid., 118.

83 For an early and influential approach to the value of "the land" and its integrity, see Aldo Leopold, *A Sand County Almanac* (Oxford: Oxford University Press, 1949).

84 Midgley, Interview by Gregory S. McElwain, March 6, 2011.

Chapter 5: Our Connection to Nature

1 Midgley, Interview by Gregory S. McElwain, March 6, 2011.

2 See C. J. Glacken, *Traces on the Rhodian Shore: Nature and Culture in Western Thought from Ancient Times to the End of the Eighteenth Century* (Berkeley: University of California Press, 1967).

3 Midgley, *The Myths We Live By*, 172.

4 See Holmes Rolston III, "Loving Nature: Christian Environmental Ethics," in *Love and Christian Ethics: Tradition, Theory, and Society*, ed. Frederick V. Simmons with Brian C. Sorrell (Washington DC: Georgetown University Press, 2016), 313–31.

5 Midgley, Interview by Gregory S. McElwain, March 6, 2011. See also Lynn White Jr., "The Historical Roots of Our Ecologic Crisis," *Science* 155, no. 3767 (1967): 1203–7.

6 For a general collection, see Dieter T. Hessel and Rosemary Radford Ruether, eds. *Christianity and Ecology: Seeking the Well-being of Earth and Humans* (Cambridge, MA: Harvard Divinity School Center for the Study of World Religions/Harvard University Press, 2000). Notable ecotheologians include Thomas Berry, John Cobb, Catherine Keller, and Sallie McFague, among others.

7 Midgley, *The Myths We Live By*, 174.

8 Augustine, *Confessions*, trans. Henry Chadwick (Oxford: Oxford University Press, 2009).

9 There is considerable debate about the relationship between religious ideas and practices. That is, whether ideas directly influence practices in a monocausal direction or whether practices likewise influence the ideas. For example, compulsory recycling may lead one to develop more positive views on the environment. Or, perhaps living close to nature in harsh conditions encourages more negative ideas of nature. In other words, ideas do not arise in a vacuum.

10 Midgley, Interview by Gregory S. McElwain, March 6, 2011.

11 See Rolston, "Loving Nature: Christian Environmental Ethics."

12 Midgley, *The Solitary Self*, 143.

13 Mary Midgley, "Introduction: The Not-So-Simple Earth," in *Earthy Realism: The Meaning of Gaia*, ed. Mary Midgley (Exeter: Imprint Academic, 2007), 5.

14 Midgley, "Beasts versus the Biosphere?" 115.

15 See Leonardo Boff, *Cry of the Earth, Cry of the Poor* (Maryknoll, NY: Orbis, 1997); Robert D. Bullard, *Dumping in Dixie: Race, Class, and Environmental Quality*, 3rd ed. (Boulder, CO: Westview Press, [1990] 2000); and Vandana Shiva, *Earth Democracy: Justice, Sustainability and Peace* (Cambridge, MA: Southend Press, 2005).

16 Global impacts on humans and nonhumans can be found in Intergovernmental Panel on Climate Change, *Climate Change 2014: Impacts, Adaptation, and Vulnerability. Part A: Global and Sectoral Aspects. Contribution of Working Group II to the Fifth Assessment Report of the Intergovernmental Panel on Climate Change* (Cambridge: Cambridge University Press, 2014) (Available online at http://www.ipcc.ch/report/ar5/wg2/).

17 Midgley, *Utopias, Dolphins and Computers*, 121–22.

18 Ibid., 124.

19 Midgley, *Evolution as a Religion*, 168. We often, furthermore, prioritize or otherwise support interests that actively endanger the environment. However much it may be in our benefit to protect the environment, the interests of the powerful can obscure our moral horizons. Political and industrial interests, for instance, can wield their power to obscure the reality of climate change by harnessing anti-science sentiments. She discusses climate change skepticism in 2011: "I've been very worried about climate change for a long time. If somebody finds a way of waking people up about climate change, that would seem to me to be of first importance. And possibly out of the present, absurd skepticism, something may emerge. I don't know. I don't see it yet. The existing sort of fear about science can easily be stirred up. And the sort of contempt for science: 'Who are these experts in their labs?' There's a really disastrous lack of sense about this. That's what is worrying me at the present. I think things may change, and it's right time they did" (Midgley, Interview by Gregory S. McElwain, March 6, 2011).

20 Midgley, *The Myths We Live By*, 11. This wider awareness is, in part, due to science, which has revealed a rich and multifaceted world full of diverse and fascinating inhabitants (Midgley, *Utopias, Dolphins and Computers*, 100). A concerned and committed scientific community has sounded the alarm, and our visions, Midgley argues, can and should respond to it. She writes that "a much saner, more realistic guiding vision for science itself does now seem to be developing out of our response to the environmental crisis. . . . Instead of joining in the celebrations of anthropolatry, they are pointing out how hard we need to work in order to protect the rest of nature from human destruction" (Mary Midgley, "Visions, Secular and Sacred," *The Hastings Center Report* 25, no. 5 [1995], 25). This, in the Copernican spirit, is what Darwin and Carl Sagan illustrate—humans are not the center of everything. Science can shine a light on the world in ways that defy our pretensions and induce marvel at the vastness of the universe. It can lead us deep into the mysteries of the world, widening our perspectives and revealing our relationality and interdependence with all things. Science, in this way, can be used to challenge our view of the world.

21 Our concepts and moral visions, Midgley argues, have been slow to catch up not only with the reality of climate change, but with the fact that we are capable of influencing global ecological health. This, she observes, "is possibly the biggest change our species has ever experienced, certainly the biggest since the invention of agriculture" (Midgley, *The Myths We Live By*, 10). The challenge, for Midgley, is that we are often at a loss when it comes to grasping the health of something as vast and holistic as the biosphere (Ibid., 246).

22 Midgley, *Utopias, Dolphins and Computers*, 130.

23 Ibid., 131–32.

24 For example, J. Baird Callicott, *In Defense of the Land Ethic: Essays in Environmental Philosophy* (Albany: State University of New York Press, 1989); Aldo Leopold, *A Sand County Almanac*; Arne Næss, "The Shallow and the Deep, Long Range Ecology Movement: A Summary," *Inquiry* 16, no. 1–4 (1973): 95–100; Rolston, *Environmental Ethics*; Taylor, "The Ethics of Respect for Nature."

25 Midgley, *Utopias, Dolphins and Computers*, 129.

26 Midgley, *The Myths We Live By*, 250–51.

27 Midgley, *The Ethical Primate*, 102–3.

28 Not to mention that our bodies themselves are miniature collectives in themselves, composed of whole ranges of microorganisms and so on. See J. Baird Callicott, *Thinking Like a Planet: The Land Ethic and the Earth Ethic* (Oxford: Oxford University Press, 2014).

29 Ibid., 178.

30 Midgley, *Evolution as a Religion*, 170.

31 Midgley, *Science and Poetry*, 20.

32 Midgley, *Utopias, Dolphins and Computers*, 124.

33 Ibid., 126.

34 Midgley, Interview by Gregory S. McElwain, March 6, 2011.

35 Donald Worster examines the extent to which organicist visions of nature influence ethics at the end of *Nature's Economy: A History of Ecological Ideas (Studies in Environment and History)*, 2nd ed. (Cambridge: Cambridge University Press, [1977] 1994). See also Glacken, *Traces on the Rhodian Shore*.

36 Midgley, *Science and Poetry*, 258. Emphasis removed.

37 Mary Midgley, "Introduction: The Not-So-Simple Earth," 3. See James Lovelock, *Gaia: A New Look at Life on Earth*, 3rd ed. (Oxford: Oxford University Press, [1979] 2000).

38 Midgley, "Visions, Secular and Sacred," 26.

39 Midgley, *Science and Poetry*, 240.

40 Gaia, of course, was a powerful and nurturing earth *goddess*. Whatever its scientific merits may or may not be, Midgley argues that envisioning feminine power in this way may perhaps be too much of a scandal for some (Ibid., 253).

41 James Lovelock, *Gaia: The Practical Science of Planetary Medicine* (Oxford: Oxford University Press, [1991] 2001).

42 Mary Midgley, Interview by Gregory S. McElwain, May 26, 2015, in Newcastle Upon Tyne, UK.

43 Mary Midgley, Interview by Gregory S. McElwain, October 23, 2017, in Newcastle Upon Tyne, UK.

44 Prominent thinkers such as J. Baird Callicott and Bruno Latour more recently have employed the metaphor of Gaia as a useful way of conceptualizing our planetary citizenship: Callicott, *Thinking Like a Planet*; Bruno Latour, *Facing Gaia: Eight Lectures on the New Climatic Regime*, trans. Catherine Porter (Cambridge: Polity Press, 2017).

45   Ibid., 185–86. This was the prominent critique of J. Baird Callicott ("Animal Liberation") and Mark Sagoff ("Animal Liberation and Environmental Ethics").

46   For example, Taylor, "The Ethics of Respect for Nature" and Rolston, *Environmental Ethics.*

47   Midgley, *Science and Poetry*, 257.

48   For Midgley, we can speak of "duties" to the nonhuman world, though the term should be used in the *everyday* sense of duties to community. Otherwise, she argues, we run the risk of getting entangled in overly technical and legalistic squabbles, which again tend to restrict and narrow our moral scope rather than widen it.

49   Midgley, "Beasts versus the Biosphere?" 117.

50   Ibid., 116.

51   Midgley, *Animals and Why They Matter*, 90.

52   Ibid., 91.

53   Midgley, "Beasts versus the Biosphere?" 115.

54   See, for instance, Carolyn Merchant, *The Death of Nature: Women, Ecology, and the Scientific Revolution* (San Francisco, CA: HarperCollins, [1980] 1990); Val Plumwood, *Feminism and the Mastery of Nature* (London: Routledge Press, 1993); and Karen Warren, "The Power and the Promise of Ecological Feminism," *Environmental Ethics* 12, no. 2 (1990): 125–46.

Chapter 6: Gender and Fragmentation

1   Midgley, Interview by Gregory S. McElwain, October 23, 2017.

2   Midgley, *Utopias, Dolphins and Computers*, 74.

3   Mary Midgley, "Sex and Personal Identity," *Encounter* 63 (1984): 50.

4   Cynthia Freeland, "Nourishing Speculation: A Feminist Reading of Aristotelian Science," in *Engendering Origins: Critical Feminist Readings in Plato and Aristotle*, ed. Bat-Ami Bar On (Albany: State University of New York Press, 1994), 145–46; Aristotle, *Generation of Animals*, trans. A. L. Peck (Cambridge, MA: Harvard University Press, 1942); Mary Midgley and Judith Hughes, *Women's Choices: Philosophical Problems Facing Feminism* (London: Weidenfeld and Nicolson, 1983), 45.

5   John Stuart Mill, *On Liberty and the Subjection of Women*, ed. Alan Ryan (New York: Penguin Classics, 2006).

6 Arthur Schopenhauer, *Parerga and Paralipomena*, Vol. II, trans. E. J. Payne (Oxford: Clarendon Press, 1974).

7 Friedrich Nietzsche, *Thus Spoke Zarathustra*, trans. Walter Kaufman (New York: Penguin Books, 1954), 178–79.

8 Midgley and Hughes, *Women's Choices*, 45.

9 Ibid.

10 Midgley, *Utopias, Dolphins and Computers*, 80; and Mary Midgley, "Rings and Books," first published online at http://www.womeninparenthesis.co.uk/wp-content/uploads/2016/05/rings-and-books.pdf.

11 Midgley and Hughes, *Women's Choices*, 39–41.

12 Immanuel Kant, *Observations on the Feeling of the Beautiful and Sublime*, trans. John T. Goldthwait (Berkeley: University of California Press, 2004).

13 Ibid., 39.

14 Ibid., 41.

15 Midgley, "Sex and Personal Identity," 53–54.

16 Midgley and Hughes, *Women's Choices*, 38.

17 Midgley, "Sex and Personal Identity," 53.

18 Ibid., 90–91.

19 Midgley, *Science as Salvation*, 50.

20 See Rosemary Radford Ruether, *Sexism and God-Talk: Toward a Feminist Theology* (Boston, MA: Beacon Press, [1983] 1993).

21 See Evelyn Fox Keller, *Reflections on Gender and Science: Tenth Anniversary Paperback Edition* (New Haven: Yale University Press, [1985] 1995); and Merchant, *The Death of Nature*.

22 Midgley, *Science as Salvation*, 88.

23 Ibid., 87–88. On Midgley's notion of a "full-time feminist," Liz McKinnell writes: "But what is a full-time feminist, in the context of philosophy, and would Midgley qualify? I would contend that the full-time feminist need not write exclusively about questions of gender, and that this might not even be a central focus of their work. Rather, they need to be dedicated to identifying and nullifying gender biases in their thought and in the thought of others, and to resist the notion that such concerns are irrelevant from a serious philosophical standpoint. On this count, as we have seen, Midgley undoubtedly qualifies. She demonstrates that even (and perhaps especially) in realms where we find explicit professions of 'neutrality' and 'objectivity,' the standpoint being adopted can be implicitly gendered" (Liz McKinnell, "On Being a 'Full-Time Feminist': Midgley and Feminist Philosophy," in *Science and the*

Self: Animals, Evolution, and Ethics: Essays in Honour of Mary Midgley, eds. Ian James Kidd and Liz McKinnell [London: Routledge, 2015], 205).

24 Ibid., 90–91.

25 Ibid., 87–88.

26 Midgley, "Introduction: The Not-So-Simple Earth," 7.

27 See Val Plumwood, *Feminism and the Mastery of Nature*; and Warren, "The Power and the Promise of Ecological Feminism."

28 Midgley, *Animals and Why They Matter*, 74.

29 Ibid., 66–67.

30 Ibid., 72. Midgley draws attention to the nineteenth-century women who fought for the rights to education and work while relying on the cheap and often exploitative labor of uneducated female servants (Midgley, *Animals and Why They Matter*, 72.). Here, class was as deep of an identifier as sex.

31 Ibid., 98.

32 Rousseau is a particularly perplexing example of inconsistent and obtuse views of women. Though he first describes women alongside men in an equally independent state of nature, he quickly relegates them to a life of passive submission, blissful ignorance, and trite vanity while being excluded from the social contract (Ibid., 76–77). Midgley argues that this tension in Rousseau and others illustrates the pathology of egalitarianism and the difficulties associated with notions of equality. In Rousseau's case, his singular life story and fear of "more formidable ladies" shaped his attitudes. Rousseau was uniquely dependent on others throughout his life, first as an orphan and servant and later under the support of patrons and especially patronesses, which clashed with his desire for freedom. Midgley argues that this contributes to the wider tendency in Rousseau and others to not fully think through the topic because of personal reasons, sloppily presenting inconsistent and ill-conceived arguments that contradict their egalitarian ideals. See McKinnell, "On Being a 'Full-time Feminist,'" 197–206.

33 Simone de Beauvoir, *The Second Sex* (New York: Vintage Books, [1952] 1989); Mary Wollstonecraft, *The Complete Works of Mary Wollstonecraft*, eds. Janet Todd and Marilyn Butler (London: William Pickering, 1989).

34 Utilizing Janet Radcliffe Richards, *The Sceptical Feminist: A Philosophical Enquiry* (London: Routledge Press, 1980).

35 Midgley and Hughes, *Women's Choices*, 24.

36 Ibid., 3.

37 Midgley, *Utopias, Dolphins and Computers*, 91; and Midgley, *Beast and Man*, xxv.

38 Midgley and Hughes, *Women's Choices*, 24.

39 Ibid., 23.

40 Ibid., 24.

41 Ibid., 5.

42 Ibid., 99.

43 Ibid., 5.

44 Ibid.

45 Midgley, *Beast and Man*, xxv.

46 Ibid., xxiv.

47 See, for instance, Claire Ainsworth, "Sex Redefined," *Nature* 518, no. 7539 (2015): 288–91, which examines the burgeoning science behind the "wider spectrum" of sex. Midgley seems to view gender as more spectral than binary, while she seems to lean toward a more binary view of sex.

48 Midgley, *Beast and Man*, 314.

49 Midgley, "Sex and Personal Identity," 52, quoting Carlyle via Tyndall.

50 Ibid., 51.

51 Ibid., 51–52.

52 Midgley and Hughes, *Women's Choices*, 10–11.

53 Ibid., 26.

54 Midgley, *Utopias, Dolphins and Computers*, 94.

55 Midgley and Hughes, *Women's Choices*, 21–22.

56 Midgley and Hughes argue that the warfare approach to gender, which was characterized in the 1970s–80s by separatism, is untenable. Coalitions of parties, including men, are required for sustainable change. Though men are, in varying measures, the source of the problem, they must be part of the solution if the world is indeed to change for the better. The rhetoric of a sex-war reinforces the same divisions and faults of patriarchal and oppositional ways of seeing the world (Ibid., 17). This cooperation is, no doubt, easier said than done. However, pragmatically, a war cannot be won by either side and, as Midgley and Hughes wryly note, it is not clear "how this could ever work for women and men, who live together and continually give each other hostages" (Ibid., 108). Midgley also notes that this cooperative approach is likely the

only way, by nature of the problem, that things will change: "The people whose attitudes need to be changed here are, by definition, men, and if they cannot be talked to—if it is only possible to preach to the converted—then the whole enterprise has to be abandoned" (Midgley, *Utopias, Dolphins and Computers*, 90).

57 Ibid., 32.

58 Ibid.

59 Ibid., 8.

60 Ibid., 69.

Chapter 7: Science in Context

1 Midgley, Interview by Gregory S. McElwain, June 11, 2014.

2 Midgley, *Evolution as a Religion*, 124.

3 Midgley, *Wisdom, Information and Wonder*, 12. Midgley highlights some challenges in the use of the term "scientific" itself. On the one hand, "scientific" can simply mean "thorough" or "methodical" (Midgley, *Science and Poetry*, 200). In this sense, a wide range of studies can be scientific. On the other hand, it is often used specifically in reference to the *scientific* disciplines (especially of those that study the natural world). In this latter case, history and disciplines outside of the natural sciences are not considered scientific. Midgley is concerned with reconciling these divergent understandings of science when appropriate.

4 Midgley argues that the historical and artificial split between empiricism and rationalism have obscured this dynamic (Midgley, *Wisdom, Information and Wonder*, 200).

5 Midgley, *Can't We Make Moral Judgements?* 18.

6 Ibid.

7 Midgley, *Science and Poetry*, 140–41 and 247.

8 Krauss's full quotation: "Philosophy is a field that, unfortunately, reminds me of that old Woody Allen joke, 'those that can't do, teach, and those that can't teach, teach gym.' And the worst part of philosophy is the philosophy of science; the only people, as far as I can tell, that read work by philosophers of science are other philosophers of science. It has no impact on physics what so ever, and I doubt that other philosophers read it because it's fairly technical. And so it's really hard to understand what justifies it. And so I'd say that this tension occurs because people in philosophy feel threatened, and they have every right to feel threatened, because science progresses and philosophy doesn't (Ross Anderson, "Has Physics Made Philosophy and Religion Obsolete?" *The Atlantic*, April 23, 2012. https://www.theatlantic.com/technology/archive/201 2/04/has-physics-made-philosophy-and-religion-obsolete/256203/ [accessed October 1, 2018]).

9 Midgley, *Science and Poetry*, 49–50. She points out that "the business of scientists is to criticize science." Midgley, Interview by Gregory S. McElwain, May 29, 2013.

10 Midgley, *Wisdom, Information and Wonder*, 117.

11 Midgley, *Science as Salvation*, 100.

12 Stephen Hawking and Leonard Mlodinow, *The Grand Design* (London: Transworld Publishers, 2010).

13 See Ian J. Kidd, "Doing Science and Injustice: Midgley on Scientism," in *Science and the Self: Animals, Evolution, and Ethics: Essays in Honour of Mary Midgley*, eds. Ian James Kidd and Liz McKinnell (London: Routledge, 2015), 151–67.

14 Midgley, *Are You an Illusion?* 5.

15 Peter Atkins, "The Limitless Power of Science," in *Nature's Imagination: The Frontiers of Scientific Vision*, ed. John Cornwell (Oxford: Oxford University Press, 1995), 123. A more tempered version that venerates the potent benefits of science is found in Jawaharlal Nehru: "Science alone can solve the problems of hunger and poverty, of insanitation and illiteracy, of superstition and deadening custom and tradition. . . . The future belongs to science and to those who make friends with science." Quoted in Midgley, *The Myths We Live By*, 14.

16 Midgley, *Evolution as a Religion*, 31.

17 Midgley, *Are You an Illusion?* 5.

18 Ibid., 5.

19 Ibid.

20 Midgley, *Science and Poetry*, 71.

21 Midgley, *Are You an Illusion?* 6.

22 Ibid. See John Dupré, *The Disorder of Things: Metaphysical Foundations of the Disunity of Science* (Cambridge, MA: Harvard University Press, 1993); and Peter Galison and David J. Stump, eds. *The Disunity of Science: Boundaries, Contexts, and Power* (Stanford, CA: Stanford University Press, 1996).

23 Midgley, *Are You an Illusion?* 6.

24 Ibid., 2.

25 Midgley, *Science and Poetry*, 74. It is important to note that, though scientists can certainly contribute to scientistic dramas—especially vocal popularizers of science—they are not typically proponents of scientistic ideas. Thus they are not the target of Midgley's critique. Though some scientists would do well to be more reflexive in their practice and dissemination of their findings, they are not to blame for the religious fervor and isolation through which their work is sometimes viewed (Midgley, *Utopias, Dolphins and Computers*, 140). Rather, sources outside of science proper, along with media sensationalism—for example, reporting of the "Holy Grail" of astronomy or neuroscience—primarily stoke the fire of scientism and perpetuate narratives of science as an almost spiritual source of guidance that rises above all other inquiry.

26 Francis Crick, *The Astonishing Hypothesis: The Scientific Search for the Soul* (New York: Touchstone, 1994), 3. Emphasis mine.

27 Midgley, *The Ethical Primate*, 69. Reductionism, Midgley argues, is not a self-evident approach to science. It should not be mistaken as value-neutral nor as tantamount to the project of science itself. It is part "of some larger enterprise, some project for reshaping the whole intellectual landscape, and often our general attitude to life as well" (Midgley, *The Myths We Live By*, 43). She also notes that, ironically, reductionist accounts do not go as far as they recommend. Their selection of fundamental explanations or units, such as genes or neurons, are largely arbitrary and matters of convenience. If reduction to the simplest physical account is the goal of inquiry, explanation of subatomic particles may be the benchmark (Midgley, *Science and Poetry*, 9–10). And, once there, the quagmire of the quantum world tends to cast doubt on any claims to fundamental explanations.

28 Midgley, *The Ethical Primate*, 55–56.

29 Midgley, *The Myths We Live By*, xiii.

30 Ibid., 39. Midgley, *Wisdom, Information and Wonder*, 192.

31 Midgley argues that metaphors are inextricably part of how we think (Midgley, *The Ethical Primate*, 87–88). On this, cf. Mary Hesse, *Models and Analogies in Science*, revised ed. (Notre Dame, IN: Notre Dame University Press, 1966); and George Lakoff and Mark Johnson, *Metaphors We Live By* (Chicago, IL: University of Chicago Press, 2003).

32 Midgley, *Science and Poetry*, 112–13.

33 Midgley, *Evolution as a Religion*, 5.

34 Midgley, *Animals and Why They Matter*, 145.

35 Midgley, *Science as Salvation*, 92.

36 Midgley, *The Myths We Live By*, 1.

37 Ibid., 19 and xii, and *Utopias, Dolphins and Computers*, 12.

38 Ibid., 4.

39 Midgley, *Evolution as a Religion*, 4–6.

40 Midgley, *The Myths We Live By*, xvii.

41 Midgley, *Science and Poetry*, 2.

42 Ibid., 102.

43 See E. Peter Greenberg, "Bacterial Communication and Group Behavior," *The Journal of Clinical Investigation* 112, no. 9 (2003): 1288–90.

44 Midgley, *The Solitary Self*, 24. Midgley notes, moreover, that atomistic visions, which are so closely associated with physical views of the world, are often out of sync with physics and rely on archaic ways of breaking up the world. Informed by quantum mechanics, physicists "have seen reason to drop this seductive vision and to recognise that the world is actually more complex" (Midgley, *Science and Poetry*, 102). Accordingly, accounts of interconnected particles are more representative of prevailing physics than isolated, indivisible, and unchangeable atoms (Ibid., 82).

45 See Midgley, *The Essential Mary Midgley*, 16; Midgley, *Evolution as a Religion*, 15.

46 Midgley, *Science as Salvation*, 38.

47 See also Midgley, *Beast and Man*, 170; and *Evolution as a Religion*, 4. Moreover, facts can be overdetermined by theory, and theory can be underdetermined by facts (See W. V. O. Quine, "Main Trends in Recent Philosophy: Two Dogmas of Empiricism," *Philosophical Review* 60, no. 1 [1951]: 20–43; and Pierre Maurice Marie Duhem, *The Aim and Structure of Physical Theory* [Princeton, NJ: Princeton University Press, 1954]). Midgley also argues that different facts require different concepts. Facts about the molecular constitution of algae require different sets of concepts than facts about feedback mechanisms in ecosystems. This does not mean these concepts clash, but rather, they interact with each other in complementary ways as part of the broader conceptual and interpretive scaffolding of scientific inquiry. Interpretive schemes are multiple and regularly revised, but their selection and implementation largely determine the questions asked and the answers found (Midgley, *The Ethical Primate*, 45–47).

48 Midgley, *Evolution as a Religion*, 15.

49 Ibid., 16.

50 Midgley, *Evolution as a Religion*, 158.

51 Ibid., 4. Darwin is, again, Midgley's exemplar here. Though his work is prone to exaggeration, misinterpretation, and misrepresentation, he made efforts to make sure that his portrayal of the natural world was balanced, and he was clear to make his own visions known. This refining process is challenging because we take our world-pictures seriously. They are vital sources of meaning and offer direction in life. Accordingly, we tend to trust or have faith in our world-pictures as given. However, if taken too seriously, one may unconsciously think they are seeing the world directly as it is, without any personal influences (Midgley, *Science as Salvation*, 60). Conversely, if not taken seriously enough, one may sink into relativism. Again, the key, for Midgley, is balance in taking one's world-picture seriously enough to employ in a meaningful way while refining it in pursuit of the best map of the world. This is most often done by comparing and checking them against others.

52 Ibid., 54.

53 Ibid., 58; and Midgley, *Science and Poetry*, 93.

54 Midgley, *Evolution as a Religion*, 26–28.

55 Albert Einstein, Personal Letter to Max Born, December 4, 1926.

56 For complications in objectivity, see Donna Haraway, "Situated Knowledges: The Science Question in Feminism and the Privilege of Partial Perspectives," *Feminist Studies* 14, no. 3 (1988): 575–99.

57 Midgley, *Science as Salvation*, 47.

58 Midgley, *Wisdom, Information and Wonder*, 49.

59 Santiago Ramón y Cajal, *Advice for a Young Investigator*, trans. Neely Swanson and Larry W. Swanson (Cambridge, MA: A Bradford Book, 1999), 89.

60 Scientific practices and communities can fall victim to wider social prejudices—for instance #MeToo in science, women and people of color not being included in clinical trials, and so on. See Sarah Kaplan and Ben Guarino, "Half of Women in Science Experience Harassment, A Sweeping New Report Finds," *Washington Post*, June 12, 2018. https://www.washingtonpost.com/news/speaking-of-sc ience/wp/2018/06/12/half-of-women-in-science-experience-harassmen t-a-sweeping-new-report-finds/ (accessed October 1, 2018). See also Robert Guthrie's analysis of racial biases and preconceptions in *Even the Rat Was White: A Historical View of Psychology*, 2nd ed. (Upper Saddle River, NJ: Pearson Education, 2004).

61 Skepticism about the claims of science often dovetails with a generalized epistemological skepticism, which questions whether we can really be confident about *any* of our knowledge (see "Morality"). This form of skepticism focuses on the scope and limitations of our cognitive faculties and is dubious that scientific knowledge can ever be perfect or final.

62 Bayesian probability theory, as an example, offers a statistical way to deal with uncertainty in the information underlying measurements. The spirit of it, rather than deciding whether a hypothesis is true or false, seeks to quantify how confident we can be that it is true.

63 Midgley, *Science as Salvation*, 120–21.

64 Midgley, *The Myth We Live By*, 40. This moderate skepticism resembles Ian Barbour's "critical realism" in Ian Barbour, *Religion in an Age of Science: The Gifford Lectures 1989-1991 Volume 1* (San Francisco, CA: Harper & Row Publishers, 1990).

Chapter 8: Religion, Science, and Complexity

1 Midgley, Interview by Gregory S. McElwain, June 11, 2014.

2 Like many other topics, Midgley argues that our fullest knowledge of religion can only come through pluralistic methods and forms of description (Midgley, *Evolution as a Religion*, 21). Though religion has philosophical elements, it is more expansive and has many features, including beliefs, practices, ethics, communities, and so on. Philosophy is one of the many lenses to studying religion, alongside history, anthropology, sociology, psychology, and so on. Religious studies, as a conglomerate and polymethodic discipline, draws on the unique insights from these distinct approaches in synthesizing the wider picture of religious phenomena in the world. This is no easy task and involves many scholars from many disciplines, each navigating the complexities of religion within the parameters and assumptions of their own approaches. Some scholars employ "insider" approaches to the study of religion—drawing on firsthand accounts of beliefs—while others employ "outsider" approaches—drawing on observations. Though there are intense debates within religious studies over whether qualitative or quantitative approaches best capture the complexity of religion, most approaches, in one way or another, utilize a sliding scale of insider and outsider approaches. Religious studies, it should also be noted, is distinct from theology.

3 Theology is, roughly, the application of philosophical methods *within* a religious tradition. The term "theology" (from the Greek *theos*, God, and *logos*, "logic/study/reasoning of"), however, is misleading, as it implies some belief in or acknowledgment of God or gods. Various traditions—for instance, versions of Buddhism and certain Chinese religions—have no close analog to such notions of personal deities. For an example of the study of lived religion, see Robert Orsi, *Between Heaven and Earth: The Religious Worlds People Make and the Scholars Who Study Them* (Princeton, NJ: Princeton University Press, 2006).

4 Midgley, *Evolution as a Religion*, 33–35. The term "Western" is fairly artificial in general, especially in relation to religions. It assumes a certain orientation or vantage point toward the world's geography while overlooking the complex historical interactions between many regions and religions. It is also notoriously vague. Here, it is used to demonstrate the preoccupation of Anglo-American and, especially, Protestant Christian approaches to religion and science. See J. Z. Smith, *Relating Religion: Essays in the Study of Religion* (Chicago, IL: University of Chicago Press, 2004).

5 See Benson Saler, "Family Resemblance and the Definition of Religion," *Historical Reflections/Réflexions Historiques* 25, no. 3 (1999): 391–404; and Victoria Harrison, "The Pragmatics of Defining Religion in a Multi-Cultural World," *International Journal for Philosophy of Religion* 59 (2006): 133–52.

6 Mary Midgley, "Visions, Secular and Sacred," 25, italics removed.

7 Midgley, *The Myths We Live By*, 1.

8 Midgley, *Evolution as a Religion*, 15.

9 Ibid., ix.

10 Cover of *Time Magazine*, November 13, 2006.

11 Barbour, *Religion in an Age of Science*.

12 Midgley, *The Myths We Live By*, 187. See "Human Nature" and "Animals."

13 This argument is made by numerous authors in Gary B. Ferngren, ed. *Science and Religion: A Historical Introduction* (Baltimore, MD: John Hopkins University Press, [2002] 2017).

14 See Pew Research Center, "The Global Religious Landscape." (Available online at http://www.pewforum.org/2012/12/18/global-rel igious-landscape-exec/).

15 Stephen Jay Gould, *Rocks of Ages: Science and Religion in the Fullness of Life* (New York: Ballantine Books, 1999); Midgley, *Evolution as a Religion*, 12–13.

16 Cf. Ferngren, *Science and Religion*. In the case of Galileo, the Catholic Church was in the midst of a larger struggle related to the authority of biblical interpretation during the Counter-Reformation (against the positions of Protestants like Martin Luther and John Calvin). Galileo, in proposing how biblical scripture should be read regarding the relative position of the earth and sun, was arguably silenced for challenging the interpretive authority of the church. He argued that truths can be found in revelation *and* in nature. If one truth seems to contradict the other, something requires reinterpretation. Since the Bible is a spiritual document and not about natural truths, geocentric verses should be reinterpreted. Geocentric scriptures, in this way, were "accommodated" by God to human understanding, but not to be understood literally.

17 Examples of creation science today include Bert Thompson, *The Scientific Case for Creation*, 3rd ed. (Montgomery, AL: Apologetics Press, [1986] 2004); and Answers in Genesis (Available online at https://answersingenesis.org). This is often a selective literalism. For instance, the dome or firmament with waters above it does not typically manifest in these accounts: "And God said, 'Let there be a dome in the midst of the waters, and let it separate the waters from the waters.' So God made the dome and separated the waters that were under the dome from the waters that were above the dome. And it was so. God called the dome Sky" (Gen. 1: 6–8). See Midgley, *Science as Salvation*, 52–53.

18 Midgley, *Evolution as a Religion*, 14.

19 Galileo Galilei, *Discoveries and Opinions of Galileo*, trans. Stillman Drake (New York: Anchor Books, 1990).

20 Ibid.

21 Gould, *Rocks of Ages*.

22 Midgley, *Evolution as a Religion*, 112.

23 Midgley, Interview by Gregory S. McElwain, June 11, 2014.

24 Midgley, *Evolution as a Religion*, 173.

25 Ibid., 1.

26 Ibid., 33.

27 Ibid., 8.

28 Ibid., 7.

29 Ibid., 38.

30 Ibid., 140.

31 Midgley argues that Social Darwinism and competitive egoism are more reflective of Hobbesian, rather than Darwinian understandings of human nature. Hobbes takes a lot of criticism from Midgley, mainly because of how influential and pervasive his ideas are today. This Hobbesian legacy lives on through various distortions of Darwin's depiction of nature. Darwin is often credited with bequeathing us the cutthroat egoism that drives such ideologies as Social Darwinism. However, his portrayal of human nature was more nuanced than this, focusing extensively on social life. Darwin's evolution of morality, for example, only makes sense in this deeply social context (see "Morality").

32 Though most distortions of science occur in scientistic literature, some scientists have presented one-sided and extreme views on human nature. These sources not only represent one-sided views of human nature and society but also distort the science itself with personal ideology, politics, and metaphysical assumptions. In these cases, the science becomes entangled with this range of factors external to science, distorting the science and leading the public to believe that their position is purely scientific. Much of the time, however, it is actually a reflection of social and political ideals transposed onto scientific description of the world (see "Science"). These positions, in their respective glorification of the human intellect (the escalator fallacy) and the forces that motivate human behavior (Social Darwinism), have, furthermore, been used for objectionable social and political uses, including eugenics.

33 Midgley, *Evolution as a Religion*, 114. Other antitheses in this group include science versus error, science versus blind conformism, and science versus childishness.

34 Ibid., 117.

35 Ibid., Other antitheses in this collection include progress versus tradition, determinism versus free will, mechanism versus teleology, empiricism versus metaphysics, quantity versus quality, physical science versus the humanities, realism versus reverence, and clarity versus mystery.

36 Ibid., 117.

37 Ibid.

38 Midgley, *The Owl of Minerva*, 19.

39 Midgley, *Evolution as a Religion*, 157.

40 Midgley, *The Solitary Self*, 94.

Chapter 9: Afterword

1 Midgley, *The Owl of Minerva*, x. In her memoir, Midgley adapts Hegel's use of the owl of Minerva, who "spreads its wings only with the falling of the dusk." She proposes that "wisdom, and therefore philosophy, comes into its own when things become dark and difficult rather than when they are clear and straightforward" (Ibid.).

2 Midgley, *Science and Poetry*, 183.

3 Midgley, Interview by Gregory S. McElwain, May 23, 2012. Originally the title of an article (Mary Midgley, "One World, But a Big One," *Journal of Consciousness Studies* 3, no. 5/6 [1996]: 500–14), she was, in our conversation, referring to the complexity of the world.

4 Midgley, *What Is Philosophy For?*

5 Ibid., 51.

6 Ibid., 208.

7 Midgley, Interview by Gregory S. McElwain, March 6, 2011.

Bibliography

Works by Midgley

* Denotes authorship as Mary Scrutton.

Books

Animals and Why They Matter. Athens, GA: University of Georgia Press, 1983.

Are You an Illusion? London: Routledge, 2014.

Beast and Man: The Roots of Human Nature, revised ed. London: Routledge, [1979] 2002.

Can't We Make Moral Judgements? London: Bloomsbury, [1989] 2017.

Evolution as a Religion: Strange Hopes and Stranger Fears. London: Routledge, [1985] 2002.

Heart and Mind: The Varieties of Moral Experience, revised ed. London: Routledge, [1981] 2003.

Science and Poetry. London: Routledge, [2001] 2006.

Science as Salvation: A Modern Myth and Its Meaning. London: Routledge, 1992.

The Essential Mary Midgley, ed. David Midgley. London: Routledge, 2005.

The Ethical Primate: Humans, Freedom and Morality. London: Routledge, 1994.

The Myths We Live By. London: Routledge, [2004] 2011.

The Owl of Minerva: A Memoir. London: Routledge, 2005.

The Solitary Self: Darwin and the Selfish Gene. Durham, NC: Acumen, 2010.

Utopias, Dolphins and Computers: Problems of Philosophical Plumbing. London: Routledge, 1996.

What Is Philosophy For? London: Bloomsbury, 2018.

Wickedness: A Philosophical Essay. London: Routledge, [1984] 2001.

Wisdom, Information and Wonder: What Is Knowledge For? London: Routledge, 1989.
(coauthored with Judith Hughes) *Women's Choices: Philosophical Problems Facing Feminism*. London: Weidenfeld and Nicolson, 1983.

Edited volumes

(ed.) *Earthy Realism: The Meaning of Gaia*. Exeter: Imprint Academic, 2007.
(ed.) *Intersubjectivity and John Ziman's Legacy*, ed. with Joan Solomon. *Journal of Consciousness Studies* 13, no. 5 (2006).

Pamphlets

Biological and Cultural Evolution: Institute for Cultural Research Monograph Series, No. 20. Tunbridge Wells: Institute for Cultural Research, 1984.
Gaia: The Next Big Idea. London: Demos, 2001.
Impact Pamphlet 15: Intelligent Design and other Ideological Problems. London: The Philosophy of Education Society of Great Britain, 2007.
"The Value of 'Useless' Research: Supporting Scholarship for the Long Run." In *Report by the Council for Science and Society*. London: The Council for Science and Society, 1989.

Articles, book chapters, and letters

"A Golden Manifesto, Part I." *Philosophy Now* 116 (2016): 34–36.
"A Golden Manifesto, Part II." *Philosophy Now* 117 (2016): 20–23.
"A Plague on Both Their Houses." *Philosophy Now* 64 (2007): 26–27.
"Addition to Fiction." *The Twentieth Century* CLIX (1956): 567–68.*
"Age Is a Continuum – Not Two Tribes to Be Made Equal." *The Guardian*, November 14, 2011.
"Alchemy Revived." *The Hastings Center Report* 30, no. 2 (2000): 41–43.
"All Too Human." *The Guardian*, December 19, 2008.
"An Intellectual Novelist." *The Listener* 1164 (1951): 993.*
"Animal Experiments." *New Scientist* 77, no. 1087 (1978): 240.
"Animal Thoughts." *New Scientist* 171, no. 2301 (2001): 71.
"Animals." *New Scientist* 78, no. 1097 (1978): 42.

"Are Families Out of Date?" (coauthored with Judith Hughes) In *Feminism and Families*, edited by Hilde Lindemann, 55–68. London: Routledge, 1997.

"Are You An Animal?" In *Animal Experimentation: The Consensus Changes*, edited by Gill Langley, 1–18. New York: Chapman & Hall, 1989.

"Art of Psychiatry." *New Scientist* 218, no. 2916 (2013): 30–31.

"Atoms, Memes, and Individuals." In *Values, Education, and the Human World*, edited by John Haldane, 234–49. Exeter: Imprint Academic, 2004.

"Beast and Man." *The Listener* 2624 (1979): 212.

"Beasts versus the Biosphere." *Environmental Values* 1, no. 1 (1992): 113–21.

"Beasts, Brutes, and Monsters." In *What Is An Animal?* edited by Tim Ingold, 35–46. London: Unwin Hyman, 1988.

"Being Objective: The Idea of Scientists as Impartial Observers Is Hard to Shake, But Is Complete Detachment Justified?" *Nature* 410, no. 753 (2001): 753.

"Being Scientific about Our Selves." *Journal of Consciousness Studies* 6, no. 4 (1999): 85–98.

"Biotechnology and Monstrosity: Why We Should Pay Attention to the 'Yuk Factor'." *The Hastings Center Report* 30, no. 5 (2000): 7–15.

"Bishop Butler: A Reply." *The Twentieth Century* CLII, no. 905 (1952): 56–62.*

"Both Nice and Nasty." *New Statesman*, March 13, 2000.

"Bourgeois Cinderellas." *The Twentieth Century* CLV (1954): 351–63.*

"Bridge-Building at Last." In *Animals and Human Society: Changing Perspectives*, edited by Aubrey Manning and James Serpell, 188–94. London: Routledge, 1994.

"Brutal Kinship." *Nature* 399, no. 6736 (1999): 537.

"Brutality and Sentimentality." *Philosophy* 54, no. 209 (1979): 385–89.

"Can Education Be Moral?" *Res Publica* 2, no. 1 (1996): 77–85.

"Can Science Save Its Soul? Some Scientists Have Begun to Talk Confidently about Understanding God and Creation. They Are Crediting Science With Power It Doesn't Possess." *New Scientist* 1832 (1992): 43–46.

"Can Specialisation Damage Your Health?" *International Journal of Moral and Social Studies* 2, no. 1 (1987).

"Choose Free Will." *New Scientist* 215, no. 2880 (2012): 30.

"Choosing the Selectors." In *The Evolution of Cultural Entities: Proceedings of the British Academy 112*, edited by Michael Wheeler, John Ziman, and Margaret A. Boden, 119–33. Oxford: Oxford University Press, 2002.

"Coals To...." *New Scientist* 84, no. 1180 (1979): 464.

"Cold Wars and Grand Conclusions." *The Guardian*, October 28, 2008.
"Consciousness, Fatalism and Science." In *The Human Person in Science and Theology*, edited by Niels Hendrik Gregerson, Willem B. Drees, and Ulf Gorman, 21–40. Edinburgh: T&T Clark, 2000.
"Consequentialism and Common Sense." *The Hastings Center Report* 10, no. 5 (1980): 43–44.
"Correspondence" (with Anthony Flew). *Journal of Applied Philosophy* 2, no. 2 (1985): 293–94.
"Counting the Cost of Revenge." *The Guardian*, June 5, 2004.
"Crimes Beyond Religion." *The Guardian*, September 30, 2006.
"Criticising the Cosmos." (with replies by Silvia Volker and Hans Radder) In *Is Nature Ever Evil? Religion, Science and Value*, edited by Willem B. Drees, 11–26. London: Routledge, 2003.
"Curiouser and Curisouser." *The Guardian*, September 27, 2003.
"Darwin's Central Problems." *Science* 268, no. 5214 (1995): 1196–98.
"Darwinism and Ethics." In *Medicine and Moral Reasoning*, edited by K. W. M. Fulford, Grant Gillett, and Janet Martin Soskice, 6–18. Cambridge: Cambridge University Press, 1994.
"Darwinism, Purpose and Meaning." In *Philosophy and Religion: Royal Institute of Philosophy Supplement 68*, edited by Anthony O'Hear, 193–201. Cambridge: Cambridge University Press, 2010.
"Dawkins Determinant." *New Scientist* 157, no. 2121 (1998): 51–52.
"Death and the Human Animal." *Philosophy Now* 89 (2012).
"De-Dramatizing Darwin." *The Monist* 67, no. 2 (1984): 200–15.
"Descartes' Prisoners." *New Statesman*, May 24, 1999.
"Designs on Darwinism." *The Guardian*, September 6, 2005.
"Determinism, Omniscience, and the Multiplicity of Explanations." *Behavioral and Brain Sciences* 22, no. 5 (1999): 900–1.
"Do We Even Act?" In *The New Brain Sciences: Perils and Prospects*, edited by Dai A. Rees and Steven Rose, 17–33. Cambridge: Cambridge University Press, 2004.
"Does Philosophy Get Out of Date?" *Philosophy Now* 103 (2014).
"Does Science Make Belief in God Obsolete? Of Course Not." *John Templeton Foundation* (2008).
"Does the Earth Concern Us?" *Gaia Circular* (2001-2002): 4–9.
"Dover Beach Revisited." In *The Oxford Handbook of Religion and Science*, edited by Philip Clayton and Zachary Simpson, 962–78. Oxford: Oxford University Press, 2006.
"Dover Beach Revisited." *Think* 4, no. 12 (2006): 69–74.
"Dover Beach: Understanding the Pains of Bereavement." *Philosophy* 81, no. 2 (2006): 209–30.
"Duties Concerning Islands." *Encounter* LX (February 1983): 36–43.
"Earth Matters: Thinking about the Environment." In *The Age of Anxiety*, edited by Sarah Dunant and Roy Porter, 59–62. London: Virago, 1996.

"Earth Song." *New Statesman*, October 2, 2000.
"Editorial Introduction." *Journal of Consciousness Studies* 13, no. 5
    (2006): 8–16.
"Elementary, Watson." *New Scientist* 196, no. 2630 (2007): 23.
"Embarrassing Relatives: Changing Perceptions of Animals." *The
    Trumpeter* 4, no. 4 (1987): 17–19.
"Enough Is Never Enough." *The Guardian*, November 28, 2002.
"Evolution as a Religion: A Comparison of Prophecies." *Zygon* 22, no.
    22 (1987): 179–94.
"Existential Issues." with Peter Hacker, Jane Heal, and Anthony O'Hear,
    New Scientist 211, no. 2829 (2011): 32.
"Fate by Fluke." *The Guardian*, March 1, 2003.
"Formulas Built in Myth." *The Guardian*, August 15, 2009.
"Freedom and Heredity, *The Listener*, September 14, 1978.
"Gene-Juggling." *Philosophy* 54, no. 210 (1979): 439–58.
"Getting Real." *New Scientist* 216, no. 2887 (2012): 30.
"Great Thinkers - James Lovelock." *New Statesman*, July 14, 2003.
"Heaven and Earth, an Awkward History." *Philosophy Now* 34 (2001–
    2002).
"Hobbes' *Leviathan*, Part 1: Strange Selves." *The Guardian*, April 6,
    2009.
"Hobbes' *Leviathan*, Part 2: Freedom and Desolation." *The Guardian*,
    April 13, 2009.
"Hobbes' *Leviathan*, Part 3: What Is Selfishness?" *The Guardian*, April
    20, 2009.
"Hobbes' *Leviathan*, Part 4: Selling Total Freedom." *The Guardian*, April
    27, 2009.
"Hobbes' *Leviathan*, Part 5: The End of Individualism." *The Guardian*,
    May 4, 2009.
"Hobbes' *Leviathan*, Part 6: Responses to Readers." *The Guardian*, May
    11, 2009.
"Hobbes' *Leviathan*, Part 7: His Idea of War." *The Guardian*, May 18,
    2009.
"Hobbes' *Leviathan*, Part 8: Can We Ride the Leviathan?" *The Guardian*,
    May 25, 2009.
"Homage to Gaia: The Life of an Independent Scientist." *Environmental
    Values* 10, no. 1 (2001): 141–42.
"Homunculus Trouble, or, What Is Applied Philosophy?" *Journal of
    Social Philosophy* 21, no. 1 (1990): 5–15.
"How Real Are You?" *Think* 1, no. 2 (2002): 35–46.
"Human Ideals and Human Needs." *Philosophy* 58, no. 223 (1983):
    89–94.
"Human Nature, Human Variety, Human Freedom." In *Being Humans:
    Anthropological Universality and Particularity in Transdisciplinary*

Perspectives, edited by Neil Roughley, 47–63. Berlin: Walter De Gruyter, 2000.

"Imagine There's No Heaven." *New Scientist* 192, no. 2572 (2006): 50–51.

"In Search of the Self: Simon Blackburn, Mary Midgley, and Colin Blakemore." *The Institute of Art and Ideas.* (Available online at https://soundcloud.com/instituteofartandideas/in-search-of-the-self.)

"Individualism and the Concept of Gaia." In *How Might We Live? Global Ethics in a New Century*, edited by Ken Booth, Tim Dunne, and Michael Cox, 29–44. Cambridge: Cambridge University Press, 2001.

"Individualism and the Concept of Gaia." *Review of International Studies* 26 (2000): 29–44.

"Intelligence Strikes." *New Scientist* 138, no. 1870 (1993): 51.

"Interview: Philosopher Mary Midgley - Thinker, Writer … and Nemesis of the Selfish Gene." *Herald Scotland*, March 31, 2015.

"Introduction: The Not-So-Simple Earth." In *Earthy Realism: The Meaning of Gaia*, edited by Mary Midgley, 3–9. Exeter: Imprint Academic, 2007.

"Is 'Moral' Dirty Word?" *Philosophy* 47, no. 181 (1972): 206–28.

"Is the Biosphere a Luxury?" *The Hastings Center Report* 22, no. 3 (1992): 7–12.

"It's All in the Mind." *The Guardian*, September 21, 2002.

"Just So Stories." *New Scientist* 163, no. 2205 (1999): 55.

"Keeping Species on Ice." In *Beyond the Bars: The Zoo Dilemma*, edited by Virginia MacKenna, Will Travers, and Jonathan Wray, 55–65. Wellingborough: Thorsons, 1987.

"Lorenz Maligned." *New Scientist*, no. 1521 (1986): 53.

"Lost in Space." *New Scientist* 211, no. 2823 (2011): 28.

"Love and Its Disappointment." *Philosophy Now* 75 (2008): 42.

"Mapping Science: in Memory of John Ziman." *Interdisciplinary Science Reviews* 30, no. 3 (2005): 195–97.

"Mary Midgley." In *Conversations on Truth*, edited by Mick Gordon and Chris Wilkinson, 142–54. London: Continuum, 2009.

"Matter of Choice." *New Scientist* 141, no. 1912 (1994): 50.

"Me And My Memes." *New Scientist* 161, no. 2179 (1999): 56.

"Metaphysics and the Limits of Science." *The Guardian*, August 28, 2010.

"Midgley on Murdoch" *The Philosophers' Magazine* 7 (1999): 45–46.

"Mind and Body: The End of Apartheid." In *Science, Consciousness and Ultimate Reality*, edited by David Lorimer, 173–97. Exeter: Imprint Academic, 2004.

"Mixed Antitheses." In *Religion and the Natural Sciences: The Range of Engagement*, edited by James E. Huchingson, 6–39. Eugene, OR: Wipf and Stock, 2005.

"Monkey Business. *The Origin of Species* Changed Man's Conception
    of Himself Forever. So Why, Asks Mary Midgley, Is Darwinism Used
    to Reinforce the Arid Individualism of Our Age?" *New Statesman*,
    September 6, 1999.
"Moral Brains." *New Scientist* 208, no. 2788 (2010): 29.
"More about Reason, Commitment and Social Anthropology."
    Philosophy 53, no. 205 (1978): 401–3.
"Must Good Causes Compete?" *Cambridge Quarterly of Healthcare
    Ethics* 2, no. 2 (1993): 133–42.
"Myths of Intellectual Isolation." *Proceedings of the Aristotelian Society*,
    LXXXIX, Part 1 (1988–89): 19–32.
"Newcastle: Comments on a Case-History." *The Twentieth Century*
    CLIX (1956): 159–68.*
"On Being an Anthrozoon: How Unique Are We?" *Minding Nature: A
    Journal of the Center for Humans and Nature* 5, no. 2 (2012): 11–16.
"On Being Reformed." *The Listener* no. 1428 (1956): 196.*
"On Being Terrestrial." In *Objectivity and Cultural Divergence: Royal
    Institute of Philosophy Lecture Series 17*, edited by S. C. Brown,
    79–91. Cambridge: Cambridge University Press, 1984.
"On Not Being Afraid of Natural Sex Differences." In *Feminist
    Perspectives in Philosophy*, edited by Morwenna Griffiths and
    Margaret Whitford, 29–41. London: Macmillan, 1988.
"On the Origin of Creationism." *New Scientist*, no. 2479 (2004): 29.
"On Trying Out One's New Sword on a Chance Wayfarer." *The Listener*,
    December 15, 1977.
"One World, But a Big One." *Journal of Consciousness Studies* 3,
    no. 5/6 (1996): 500–14.
"Open Letter." *Philosophy* 63, no. (1988): 1–2.
"Ou Sont les Neiges de ma Tante." *The Twentieth Century* (1959):
    168–79.*
"Paradoxography." *The Listener* no. 1183 (1951): 743.*
"Persons and Non-Persons." In *In Defence of Animals*, edited by Peter
    Singer, 52–62. Oxford: Blackwell, 1985.
"Philosophical Plumbing." In *The Impulse to Philosophise: Royal
    Institute of Philosophy Supplement 33*, edited by Phillips Griffiths,
    139–51. Cambridge: Cambridge University Press, 1992.
"Philosophising Out in the World." *Social Research* 52, no. 3 (1985):
    447–70.
"Pi in the Sky." *Third Way* 19, no. 2 (1996): 11–14.
"Pluralism: The Many Maps Model." *Philosophy Now* 35 (2002): 10–11.
"Practical Solutions." *The Hastings Center Report* 19, no. 6 (1989):
    44–45.
"Proud Not to Be a Doctor." *The Guardian*, October 3, 2005.
"Purpose, Meaning and Darwinism." *Philosophy Now* 71 (2009): 16–19.

"Putting Ourselves Together Again." In *Consciousness and Human Identity*, edited by John Cornwall. Oxford: Oxford University Press, 1998.

"Reason Is Just another Faith." *New Scientist* 199, no. 2666 (2008): 50–51.

"Reductive Megalomania." In *Nature's Imagination: The Frontiers of Scientific Vision*, edited by John Cornwell, 133–47. Oxford: Oxford University Press, 1995.

"Reductivism, Fatalism, and Sociobiology." *Journal of Applied Philosophy* 1, no. 1 (1984): 107–14.

"Religious Agenda." *New Scientist* 201, no. 2700 (2009): 22–23.

"Reply to Marc Bekoff." *Environmental Values* 1, no. 3 (1992): 256.

"Reply to Mr. Pratt." *Philosophical Books* 21, no. 1 (1980): 6–9.

"Reply to Target Article: "Inventing the Subject; the Renewal of 'Psychological' Psychology." *Journal of Anthropological Psychology* 11 (2002): 44–45.

"Rethinking Sex and the Selfish Gene: Why We Do It." *Heredity* 93, no. 3 (2006): 271–72.

"Rights Talk Will Not Sort Out Child Abuse: Comment on Archard on Parental Rights." *Journal of Applied Philosophy* 8, no. 1 (1991): 103–14.

"Rings and Books." First published online at http://www.womeninpa renthesis.co.uk/wp-content/uploads/2016/05/rings-and-books.pdf.

"Science and Poetry." *Situation Analysis* 2 (2003): 29–31.

"Science and Poetry." In *Values, Education, and the Human World*, edited by John Haldane, 219–33. Exeter: Imprint Academic, 2004.

"Science in the World." *Science Studies* 9, no. 2 (1996): 49–58.

"Science, Scientism, and the Self." *Talk for the Ian Ramsey Centre for the Humane Philosophy Project*. (Available online at https://www.you tube.com/watch?v=ekp1hEjtPsQ.)

"Scientific Whaling." *New Scientist* no. 1569 (1987): 73.

"Scientism and Free-Market Jihad." *openDemocracy*, November 19, 2014. (Available online at https://www.opendemocracy.net/mary-midgley/scientism-and-freemarket-jihad.)

"Selectionism Can only Take Us so Far." *The Guardian*, February 9, 2009.

"Selfish Genes." *New Scientist* 81, no. 1139 (1979): 278.

"Selfish Genes and Social Darwinian." *Philosophy* 58, no. 225 (1983): 365–77.

"Should We Let Them Go?" In *Attitudes to Animals: Views in Animal Welfare*, edited by Francine L. Dolins, 152–63. Cambridge: Cambridge University Press, 1999.

"Skimpole Unmasked." *History of the Human Sciences* 10, no. 4 (1997): 92–96.

"Sneering at Philosophy." *New Scientist* 98, no. 1363 (1983): 887.
"Sociobiology." *Journal of Medical Ethics* 10 (1984): 158–60.
"Sorting Out the Zeitgeist." *Changing English: Studies in Culture and Education* 7, no. 1 (2000): 89–92.
"Souls, Minds, Bodies, and Planets (part 1)." *Philosophy Now* 47 (2004).
"Souls, Minds, Bodies, and Planets (part 2)." *Philosophy Now* 48 (2004).
"Souls, Minds, Bodies, and Planets." In *Philosophy, Biology, and Life: Royal Institute of Philosophy Supplement 56*, edited by Anthony O'Hear, 83–104. Cambridge: Cambridge University Press, 2005.
"Strange Contest, Science versus Religion." In *The Gospel and Contemporary Culture*, edited by Hugh Montefiore, 40–57. London: Mowbray, 1992.
"Sustainability and Moral Pluralism." *Ethics and the Environment* 1, no. 1 (1996): 41–54.
"Teleological Theories of Morality." In *An Encyclopedia of Philosophy*, edited by G. H. R. Parkinson, 541–67. London: Routledge, 1988.
"The Absence of a Gap between Facts and Values." Co-authored with Stephen R. L. Clark. *Aristotelian Society Supplementary Volume* 54, (1980): 207–23.
"The Abuses of Science." *The Guardian*, June 12, 2010.
"The All-Female Number." *Philosophy* 54, no. 210 (1979): 552–54.
"'The Bankers' Abstract Vision of the Globe Is Limited." *The Guardian*, October 4, 2001.
"The Challenge of Science: Limited Knowledge, or a New High Priesthood?" In *True to This Earth: Global Challenges and Transforming Faith*, edited by Alan Race and Roger Williamson, 75–84. Oxford: Oneworld Publications, 1995.
"The Concept of Beastliness." *Philosophy* 48, no. 148 (1973): 111–35.
"The Emancipation of Women." *The Twentieth Century* CLII, no. 901 (1952): 217–25.*
"The End of Anthropocentrism?" In *Philosophy and the Natural Environment: Royal Institute of Philosophy Supplement 36*, edited by Robin Attfield and Andrew Belsey, 103–12. Cambridge: Cambridge University Press, 1994.
"The Ethical Primate. Anthony Freeman in Discussion with Mary Midgley." *Journal of Consciousness Studies* 2, no. 1 (1995): 67–75.
"The Flawed Frontier." *New Scientist* 195, no. 2623 (2007): 22.
"The Flight From Blame." *Philosophy* 62, no. 241 (1987): 271–91.
"The Four-Leggeds, The Two-Leggeds and the Wingeds." *Society and Animals* 1, no. 1 (1993): 9–15.
"The Game Game." *Philosophy* 49, no. 189 (1974): 231–53.
"The Idea of Salvation Through Science." *New Blackfriars* 73, no. 860 (1992): 257–65.

"The Mixed Community." In *Earth Ethics, Environmental Ethics, Animal Rights and Practical Applications*, edited by James P. Sterba, 80–90. New Jersey: Prentice Hall, 1995.

"The Month." *The Twentieth Century* CLXV (1959): 505–10.*

"The Mythology of Selfishness." *The Philosophers' Magazine* 53 (2011): 35–45.

"The Myths We Live By." In *The Values of Science: The Oxford Amnesty Lectures 1997*, Wes Williams. New York: Westview Press, 1999.

"The Natural History of Contradictions." *The Listener* issue 1180 (1951): 589.*

"The Need for Wonder." In *God For the 21st Century*, edited by Russell Stannard, 186–88. Radnot: Templeton Foundation Press, 2000.

"The Neutrality of the Moral Philosopher." *Supplementary Volume of the Aristotelian Society for 1974* (1974): 211–29.

"The Objection to Systematic Humbug." *Philosophy* 53, no. 204 (1978): 147–69.

"The Origin of Ethics." In *A Companion to Ethics*, edited by Peter Singer, 3–13. Oxford: Blackwell, 1991.

"The Origins of Don Giovanni." *Philosophy Now* 25 (1999–2000): 32.

"The Problem of Humbug." In *Media Ethics: A Philosophical Approach*, edited by Matthew Kieran, 37–48. London: Routledge, 1998.

"The Problem of Living with Wildness." In *Wolves and Human Communities, Biology, Politics and Ethics*, edited by Virginia A. Sharpe, Bryan Norton, and Strachan Donelley, 179–90. Washington DC: Island Press, 2001.

"The Problem of Natural Evil." In *Philosophy of Religion: An Anthology*, edited by Charles Talliaferro and Paul J. Griffiths, 361–67. Oxford: Blackwell, 2003.

"The Refractory: Vision." *The Lancet* 357, no. 9266 (2001): 1455.

"The Selfish Metaphor: Conceits of Evolution." *New Scientist* 209, no. 2797 (2011): 26–27.

"The Significance of Species." In *The Moral Life*, edited by Steven Luper-Foy and Curtis Brown Harcourt Brace Jovanovich, 1992.

"The Soul's Successors: Philosophy and the 'Body.'" In *Religion and the Body*, edited by Sarah Coakley, 53–70. Cambridge: Cambridge University Press, 1997.

"The Use and Uselessness of Learning." *European Journal of Education* 25, no. 3 (1990): 283–94.

"The View from Britain: What Is Dissolving Families?" (coauthored with Judith Hughes), *American Philosophical Association, Newsletter on Feminism and Philosophy* 96, no. 1 (1996).

"The Woman's Point of View." *The Listener* issue 1204 (1952): 510.*

"There Are Truths Far Too Big to Be Conveyed in One Go." *The Guardian*, December 27, 2010.

"Thinking Matter." *New Scientist* 201, no. 2688 (2008): 16.
"Towards a More Humane View of the Beasts?" In *The Environment in Question*, edited by David E. Cooper and Joy A. Palmer, 28–36. London: Routledge, 1992.
"Towards a New Understanding of Human Nature: The Limits of Individualism." In *How Humans Adapt: A Biocultural Odyssey*, edited by Donald J. Ortner, 517–46. Washington DC: Smithsonian Press, 1983.
"Towards an Ethic of Global Responsibility." In *Human Rights in Global Politics*, edited by Tim Dunne and Nicholas J. Wheeler, 160–74. Cambridge: Cambridge University Press, 1999.
"Tribal Warfare." *New Scientist* no. 1583 (1987): 73.
"Trouble with Families?" (coauthored with Judith Hughes) In *Introducing Applied Ethics*, edited by Brenda Almond, 17–32. Oxford: Blackwell, 1995.
"Two Become One." *New Scientist* 217, no. 2909 (2013): 30.
"Understanding the 'War on Terrorism.'" *openDemocracy*, October 24, 2002.
"Us and Them." *New Statesman*, September 13, 2004.
"Visions and Values." *Resurgence* 228 (2005): 18.
"Visions of Embattled Science." In *Science Today: Problem or Crisis?* edited by Ralph Levinson and Jeff Thomas, 35–50. London: Routledge, 1997.
"Visions, Secular and Sacred." *The Hastings Center Report* 25, no. 5 (1995): 20–27.
"Walk on the Dark Side." *The Philosophers' Magazine* 14 (2001): 23–25.
"Ways of Seeing." *New Scientist* 192, no. 2578 (2006): 25.
"What Darwin Got Wrong." *The Guardian*, February 6, 2010.
"What Do We Mean By Security?" *Philosophy Now* 61 (2007): 12–15.
"Why Farm Animals Matter." In *The Future of Animal Farming: Renewing the Ancient Contract*, edited by Marian Stamp Dawkins and Roland Bonney, 21–32. Oxford: Blackwell, 2008.
"Why Knowledge Matters." In *Animals in Research: New Perspectives in Animal Experimentation*, edited by David Sperlinger, 319–36. Oxford: John Wiley and Sons, 1981.
"Why Memes?" In *Alas, Poor Darwin: Arguments against Evolutionary Psychology*, edited by Hilary Rose and Steven Rose, 67–84. London: Jonathan Cape Rosenthal, 2000.
"Why Smartness Is Not Enough." In *Rethinking the Curriculum; Towards an Integrated, Interdisciplinary College Education*, edited by Mary E. Clark and Sandra A. Wawritko, 39–52. New York: Greenwood Press, 1990.
"Why The Idea of Purpose Won't Go Away." *Philosophy* 86 (2011): 545–61.

"Zombies and the Turing Test." *Journal of Consciousness Studies* 2, no. 4 (1995): 351–52.
"Zombies Can't Concentrate." *Philosophy Now* 44 (2004): 13–14.

Forewords

"Foreword" to *Creation, Environment and Ethics*, edited by Rebekah Humphreys and Sophie Vlacos, vii–iii. Newcastle upon Tyne: Cambridge Scholars Publishing, 2010.
"Foreword" to Eric Walter Frederick Tomlin, *Psyche, Culture, and the New Science; The Role of PN*, viii. London: Routledge and Kegan Paul, 1985.
"Foreword" to Georg Breuer, *Sociobiology and the Human Dimension*, ix–xii. Cambridge: Cambridge University Press, 1983.
"Foreword" to Iris Murdoch, *The Sovereignty of Good*, x–xvi. London: Routledge, 2014.
"Foreword" to M. O'Brien, *Dialogue and the Discursive Character of the Academic Discipline: Rethinking Widening Participation from the 'Inside-out.'* Liverpool: University of Liverpool, 2008.
"Foreword" to Mark Thornton, *Do We Have Free Will?* Bristol: Bristol Classical Press, 1989.
"Foreword" to Michael Bavidge, *Mad or Bad?* Bristol: Bristol Classical Press, 1989.
"Foreword" to *The Edge of Reason? Science and Religion in Modern Society*, edited by Alex Bentley, xvii–xxiv. London: Continuum, 2008.

Works by others

Ainsworth, Claire. "Sex Redefined." *Nature* 518, no. 7539 (2015): 288–91.
Anderson, Ross. "Has Physics Made Philosophy and Religion Obsolete?" *The Atlantic*, April 23, 2012. https://www.theatlantic.com/technology/archive/2012/04/has-physics-made-philosophy-and-religion-obsolete/256203/ (accessed October 1, 2018).
Anscombe, G. E. M. "Modern Moral Philosophy." *Philosophy* 33 no. 124 (1958): 1–19.
Aristotle. *Generation of Animals*, translated by A. L. Peck. Cambridge, MA: Harvard University Press, 1942.
Arluke, Arnold, Jack Levin, Carter Luke, and Frank Ascione. "The Relationship of Animal Abuse to Violence and Other Forms of

Antisocial Behavior." *Journal of Interpersonal Violence* 14, no. 9 (1999): 963–75.

Atkins, Peter. "The Limitless Power of Science." In *Nature's Imagination: The Frontiers of Scientific Vision*, edited by John Cornwell. Oxford: Oxford University Press, 1995.

Augustine, *Confessions*, translated by Henry Chadwick. Oxford: Oxford University Press, 2009.

Barbour, Ian. *Religion in an Age of Science: The Gifford Lectures 1989-1991 Volume 1*. San Francisco, CA: Harper & Row Publishers, 1990.

Beauvoir, Simone de. *Force of Circumstance, Vol. I: After the War, 1944-1952; Vol. 2: Hard Times, 1952-1962*, translated by Richard Howard. New York: Paragon House, 1992.

Beauvoir, Simone de. *The Second Sex*. New York: Vintage Books, [1952] 1989.

Bekoff, Marc. *Minding Animals: Awareness, Emotions, and Heart*. Oxford: Oxford University Press, 2003.

Berlin, Isaiah. *Four Essays on Liberty*. Oxford: Oxford University Press, 1990.

Boff, Leonardo. *Cry of the Earth, Cry of the Poor*. Maryknoll, NY: Orbis, 1997.

Bullard, Robert D. *Dumping in Dixie: Race, Class, and Environmental Quality*, 3rd ed. Boulder, CO: Westview Press, [1990] 2000.

Callicott, J. Baird. "Animal Liberation: A Triangular Affair." *Environmental Ethics* 2, no 4. (1980): 311–38.

Callicott, J. Baird. *In Defense of the Land Ethic: Essays in Environmental Philosophy*. Albany, NY: State University of New York Press, 1989.

Callicott, J. Baird. *Thinking Like a Planet: The Land Ethic and the Earth Ethic*. Oxford: Oxford University Press, 2014.

Clark, Stephen. *The Moral Status of Animals*. Oxford: Clarendon Press, 1977.

Clement, Grace. "The Ethic of Care and the Problem of Wild Animals." *Between the Species* 3 (2003).

Crick, Francis. *The Astonishing Hypothesis: The Scientific Search for the Soul*. New York: Touchstone, 1994.

Cooper, David. *Animals and Misanthropy*. London: Routledge, 2018.

Darwin, Charles. *The Descent of Man*. Princeton, NJ: Princeton University Press, [1871] 1981.

Dawkins, Richard. *The Selfish Gene*. Oxford: Oxford University Press, 1976.

Descartes, René. *Discourse on the Method*, translated by Donald A. Cress. Indianapolis: Hackett Publishing, [1637] 1998.

Descartes, René. *Meditations on First Philosophy: With Selections from the Objections and Replies*, translated by John Cottingham. Cambridge: Cambridge University Press [1986] 2017.

Duhem, Pierre Maurice Marie. *The Aim and Structure of Physical Theory*. Princeton, NJ: Princeton University Press, 1954.

Dupré, John. *The Disorder of Things. Metaphysical Foundations of the Disunity of Science*. Cambridge, MA: Harvard University Press, 1993.

Einstein, Albert. Personal Letter to Max Born, December 4, 1926.

Ferngren, Gary B., ed. *Science and Religion: A Historical Introduction*. Baltimore, MD: John Hopkins University Press, 2002.

Foot, Philippa. *Natural Goodness*. Oxford: Clarendon Press, 2003.

Foot, Philippa. "Moral Beliefs." *Proceedings of the Aristotelian Society* 59 (1958): 83–104.

Foots, Philippa. "The Philosopher's Defence of Morality." *Philosophy* 27, no. 103 (1952): 311–28.

Freeland, Cynthia. "Nourishing Speculation: A Feminist Reading of Aristotelian Science." In *Engendering Origins: Critical Feminist Readings in Plato and Aristotle*, edited by Bat-Ami Bar On. Albany: State University of New York Press, 1994.

Frey, R. G. *Interests and Rights: The Case against Animals*. Oxford: Clarendon Press, 1980.

Galilei, Galileo. *Discoveries and Opinions of Galileo*, translated by Stillman Drake. New York: Anchor Books, 1990.

Galison, Peter and David J. Stump, eds. *The Disunity of Science: Boundaries, Contexts, and Power*. Stanford, CA: Stanford University Press, 1996.

Geertz, Clifford. *The Interpretation of Cultures: Selected Essays*. New York: Basic Books, [1973] 1977.

Glacken, C. J. *Traces on the Rhodian Shore: Nature and Culture in Western Thought from Ancient Times to the End of the Eighteenth Century*. Berkeley: University of California Press, 1967.

Goodall, Jane. "Foreword." In *Science and the Self: Animals, Evolution, and Ethics: Essays in Honour of Mary Midgley*, edited by Ian James Kidd and Liz McKinnell. London: Routledge, 2015.

Gould, Stephen Jay. *Rock of Ages: Science and Religion in the Fullness of Life*. New York: Ballantine Books, 1999.

Greenberg, E. Peter. "Bacterial Communication and Group Behavior." *The Journal of Clinical Investigation* 112, no. 9 (2003): 1288–90.

Ground, Ian. "Minding Animals." In *Science and the Self: Animals, Evolution, and Ethics: Essays in Honour of Mary Midgley*, edited by Ian James Kidd and Liz McKinnell, 129–48. London: Routledge, 2015.

Gustafson, James M. *Ethics from a Theocentric Perspective*, Vols. 1–2. Chicago, IL: University of Chicago Press, 1983.

Guthrie, Robert. *Even the Rat Was White: A Historical View of Psychology*, 2nd ed. Upper Saddle River, NJ: Pearson Education, 2004.

Hadley, John. "Critique of Callicott's Biosocial Moral Theory." *Ethics & the Environment* 12, no. 1 (2007): 67–78.

Haraway, Donna. "Situated Knowledges: The Science Question in Feminism and the Privilege of Partial Perspectives." *Feminist Studies* 14, no. 3 (1988): 575–99.

Hare, R. M. *Freedom and Reason*. Oxford: Oxford University Press, 1963.

Hardin, Garrett. "Lifeboat Ethics: The Case against Helping the Poor." *Psychology Today* September (1974): 800–12.

Harrison, Victoria. "The Pragmatics of Defining Religion in a Multi-Cultural World." *International Journal for Philosophy of Religion* 59 (2006): 133–52.

Hawking, Stephen and Leonard Mlodinow. *The Grand Design*. London: Transworld Publishers, 2010.

Hesse, Mary. *Models and Analogies in Science*, revised ed. Notre Dame, IN: Notre Dame University Press, 1966.

Hessel, Dieter T. and Rosemary Radford Ruether, eds. *Christianity and Ecology: Seeking the Well-Being of Earth and Humans*. Cambridge, MA: Harvard Divinity School Center for the Study of World Religions/Harvard University Press, 2000.

Hobbes, Thomas. *Leviathan: With Selected Variants from the Latin Edition of 1668*, edited by E. M. Curley. Indianapolis, IN: Hackett, 1994.

Huxley, T. H. "On the Hypothesis That Animals Are Automata, and Its History." *Fortnightly Review* 95 (1874): 555–80.

Intergovernmental Panel on Climate Change. *Climate Change 2014: Impacts, Adaptation, and Vulnerability. Part A: Global and Sectoral Aspects. Contribution of Working Group II to the Fifth Assessment Report of the Intergovernmental Panel on Climate Change*. Cambridge: Cambridge University Press, 2014. (Available online at http://www.ipcc.ch/report/ar5/wg2/.)

James, Williams. *The Principles of Psychology, Vol. 2*. New York: Dover, [1890] 1950.

Kant, Immanuel. *Lectures on Ethics*, translated by Louis Infield. Indianapolis, IN: Hackett Publishing, 1980.

Kant, Immanuel. *Observations on the Feeling of the Beautiful and Sublime*, translated by John T. Goldthwait. Berkeley, CA: University of California Press, 2004.

Kaplan, Sarah and Ben Guarino. "Half of Women in Science Experience Harassment, A Sweeping New Report Finds." *Washington Post*, June 12, 2018. https://www.washingtonpost.com/news/speaking-of-science/wp/2018/06/12/half-of-women-in-science-experience-harassment-a-sweeping-new-report-finds/ (accessed October 1, 2018).

Keller, Evelyn Fox. *Reflections on Gender and Science: Tenth Anniversary Paperback Edition*. New Haven, CT: Yale University Press, [1985] 1995.

Kidd, Ian James "Doing Science and Injustice: Midgley on Scientism."
  In *Science and the Self: Animals, Evolution, and Ethics: Essays
  in Honour of Mary Midgley*, edited by Ian James Kidd and Liz
  McKinnell, 151–67. London: Routledge, 2015.
Kidd, Ian James and Liz McKinnell. "Introduction." In *Science and the
  Self: Animals, Evolution, and Ethics: Essays in Honour of Mary
  Midgley*, edited by Ian James Kidd and Liz McKinnell, 1–15. London:
  Routledge, 2015.
Kidd, Ian James and Liz McKinnell. *Science and the Self: Animals,
  Evolution, and Ethics: Essays in Honour of Mary Midgley.* London:
  Routledge, 2015.
King, Barbara. *Personalities on the Plate: The Lives and Minds of
  Animals We Eat.* Chicago, IL: University of Chicago Press, 2017.
Lakoff, George and Mark Johnson. *Metaphors We Live By.* Chicago, IL:
  University of Chicago Press, 2003.
Latour, Bruno. *Facing Gaia: Eight Lectures on the New Climatic Regime*,
  translated by Catherine Porter. Cambridge: Polity Press, 2017.
Leopold, Aldo. *A Sand County Almanac.* Oxford: Oxford University
  Press, 1949.
Leahy, Michael. *Against Liberation: Putting Animals in Perspective.*
  London: Routledge, [1991] 1994.
Lents, Nathan H. *Not So Different: Finding Human Nature in Animals.*
  New York: Columbia University Press, 2016.
Linzey, Andrew. *Why Animal Suffering Matters.* Oxford: Oxford
  University Press, 2009.
Lipscomb, Benjamin. "'Slipping Out Over the Wall': Midgley, Anscombe,
  Foot and Murdoch." In *Science and the Self: Animals, Evolution, and
  Ethics: Essays in Honour of Mary Midgley*, edited by Ian James Kidd
  and Liz McKinnell, 207–23. London: Routledge, 2015.
Locke, John. *An Essay Concerning Human Understanding*, edited by
  Peter H. Nidditch. Oxford: Clarendon Press, 1979.
Locke, John. *Two Treatises of Government*, edited by P. Laslett.
  Cambridge: Cambridge University Press, [1690] 1967.
Lovelock, James. *Gaia: A New Look at Life on Earth*, 3rd ed. Oxford:
  Oxford University Press, [1979] 2000.
Lovelock, James. *Gaia: The Practical Science of Planetary Medicine.*
  Oxford: Oxford University Press, [1991] 2001.
McElwain, Gregory S. "Midgley at the Intersection of Animal and
  Environmental Ethics." *Les Ateliers de l'Éthique / the Ethics Forum*
  13, no. 1 (2018):143–58.
McElwain, Gregory S. "The Mixed Community." In *Science and the
  Self: Animals, Evolution, and Ethics: Essays in Honour of Mary
  Midgley*, edited by Ian James Kidd and Liz McKinnell, 41–51. London:
  Routledge, 2015.

McKinnell, Liz. "On Being a 'Full-time Feminist': Midgley and Feminist Philosophy." In *Science and the Self: Animals, Evolution, and Ethics: Essays in Honour of Mary Midgley*, edited by Ian James Kidd and Liz McKinnell, 197–206. London: Routledge, 2015.

Merchant, Carolyn. *The Death of Nature: Women, Ecology, and the Scientific Revolution*. San Francisco, CA: HarperCollins, [1980] 1990.

Midgley, Mary. Interviews by Gregory S. McElwain. 2011–2018. Newcastle Upon Tyne, UK.

Mill, John Stuart. *On Liberty and the Subjection of Women*, edited by Alan Ryan. New York: Penguin Classics, 2006.

Moore, G. E. *Principia Ethica*. Cambridge: Cambridge University Press, [1903] 1948.

Murdoch, Iris. *The Sovereignty of Good*. New York: Schocken Books, 1970.

Næss, Arne. "The Shallow and the Deep, Long-Range Ecology Movement. A Summary." *Inquiry* 16, no. 1–4 (1973): 95–100.

Nietzsche, Friedrich. *Thus Spoke Zarathustra*, translated by Walter Kaufman. New York: Penguin Books, 1954.

Orsi, Robert. *Between Heaven and Earth: The Religious Worlds People Make and the Scholars Who Study Them*. Princeton, NJ: Princeton University Press, 2006.

Palmer, Clare. *Animal Ethics in Context*. New York: Columbia University Press, 2010.

Palmer, Clare. "What (If Anything) Do We Owe to Wild Animals?" *Between the Species* 16, no. 1 (2013): 15–38.

Polkinghorne, John. *Science and Religion in Quest of Truth*. New Haven, CT: Yale University Press, 2011.

Peterson, Anna L. *Being Animal: Beasts and Boundaries in Nature Ethics*. New York: Columbia University Press, 2013.

Plumwood, Val. *Feminism and the Mastery of Nature*. London: Routledge Press, 1993.

Quine, W. V. O. "Main Trends in Recent Philosophy: Two Dogmas of Empiricism." *Philosophical Review* 60, no. 1 (1951): 20–43.

Rachels, James. *Created from Animals: The Moral Implications of Darwinism*. Oxford: Oxford University Press, 1990.

Rawls, John. *A Theory of Justice*. Cambridge, MA: Belknap Press, [1971] 2005.

Regan, Tom. *The Case for Animal Rights*. Berkeley, CA: University of California Press, 1983.

Richards, Janet Radcliffe. *The Sceptical Feminist: A Philosophical Enquiry*. London: Routledge Press, 1980.

Rolston III, Holmes. *Environmental Ethics: Duties to and Values in The Natural World*. Philadelphia, PA: Temple University Press, 1989.

Rolston III, Holmes. "Loving Nature: Christian Environmental Ethics." In *Love and Christian Ethics: Tradition, Theory, and Society*, edited by

Frederick V. Simmons with Brian C. Sorrell, 313–31. Washington DC: Georgetown University Press, 2016.

Rousseau, Jean-Jacques. *'The Social Contract' and Other Later Political Writings*, translated by Victor Gourevitch. Cambridge: Cambridge University Press, 1997.

Ruether, Rosemary Radford. *Sexism and God-Talk: Toward a Feminist Theology*. Boston, MA: Beacon Press, [1983] 1993.

Ryle, Gilbert. *The Concept of Mind: 60th Anniversary Edition*. London: Routledge, [1949] 2009.

Sagoff, Mark. "Animal Liberation and Environmental Ethics: Bad Marriage, Quick Divorce." *Osgoode Hall Law Journal* 22, no. 2 (1984): 297–307.

Saler, Benson. "Family Resemblance and the Definition of Religion." *Historical Reflections/Réflexions Historiques* 25, no. 3 (1999): 391–404.

Santiago y Cajal, Ramón. *Advice for a Young Investigator*, translated by Neely Swanson and Larry W. Swanson. Cambridge, MA: A Bradford Book, 1999.

Sartre, Jean-Paul. "Existentialism Is a Humanism." In *The Continental Ethics Reader*, edited by Matthew Calarco and Peter Atterton, 89–94. London: Routledge, 2003.

Serpell, James. *In the Company of Animals*. Cambridge: Cambridge University Press, [1986] 1996.

Shiva, Vandana. *Earth Democracy: Justice, Sustainability and Peace*. Cambridge, MA: Southend Press, 2005.

Singer, Peter. *Animal Liberation*, 2nd ed. New York: New York Review of Books, [1976] 1990.

Skinner, B. F. *Science and Human Behavior*. New York: The Free Press [1953] 1965.

Smith, J. Z. *Relating Religion: Essays in the Study of Religion*. Chicago, IL: University of Chicago Press, 2004.

Spencer, Herbert. *Principles of Biology, Volume I*. New York: Appleton, [1864] 1898.

Spinoza, Baruch. *Ethics*, translated by E. M. Curley. Princeton, NJ: Penguin, [1677] 2005.

Taylor, Paul. "The Ethics of Respect for Nature." *Environmental Ethics* 3, no. 3 (1981): 197–218.

Thompson, Bert. *The Scientific Case for Creation*, 3rd ed. Montgomery, AL: Apologetics Press, [1986] 2004.

Voltaire. *The Philosophical Dictionary*, translated by. H. I. Woolf. New York: A.A. Knopf, 1924. (Available online at https://history.hanover.edu/texts/voltaire/volanima.html.)

Waal, Frans de. *Are We Smart Enough to Know How Smart Animals Are?* New York: W.W. Norton & Company, 2016.

Waal, Frans de. *Mama's Last Hug: Animal Emotions and What They Tell Us about Ourselves*. New York: W.W. Norton & Company, 2019.

Warnock, Mary. "Is There Such a Thing as Consensus Morality?," In *Science and the Self: Animals, Evolution, and Ethics: Essays in Honour of Mary Midgley*, edited by Ian James Kidd and Liz McKinnell, 85–99. London: Routledge, 2015.

Warren, Karen. "The Power and the Promise of Ecological Feminism." *Environmental Ethics* 12, no. 2 (1990): 125–46.

White Jr., Lynn. "The Historical Roots of Our Ecologic Crisis." *Science* 155, no. 3767 (1967): 1203–7.

Williams, Raymond. *Problems in Materialism and Culture: Selected Essays*. London: Verso Classics, 1997.

Wilson, Edward O. *Sociobiology: The New Synthesis*. Cambridge, MA: Belknap Press of Harvard University Press, [1975] 2000.

Wollstonecraft, Mary. *The Complete Works of Mary Wollstonecraft*, edited by Janet Todd and Marilyn Butler. London: William Pickering, 1989.

Worster, Donald. *Nature's Economy: A History of Ecological Ideas (Studies in Environment and History)*, 2nd ed. Cambridge: Cambridge University Press, [1977] 1994.

Index